Nutshell Series

of

WEST PUBLISHING COMPANY

P.O. Box 43526

St. Paul, Minnesota 55164

January, 1984

Accounting—Law and, 1984, approximately 392 pages, by E. McGruder Faris, Professor of Law, Stetson University.

Administrative Law and Process, 2nd Ed., 1981, 445 pages, by Ernest Gellhorn, Dean and Professor of Law, Case Western Reserve University and Barry B. Boyer, Professor of Law, SUNY, Buffalo.

Admiralty, 1983, 400 pages, by Frank L. Maraist, Professor of Law, Louisiana State University.

Agency-Partnership, 1977, 364 pages, by Roscoe T. Steffen, Late Professor of Law, University of Chicago.

American Indian Law, 1981, 288 pages, by William C. Canby, Jr., former Professor of Law, Arizona State University.

Antitrust Law and Economics, 2nd Ed., 1981, 425 pages, by Ernest Gellhorn, Dean and Professor of Law, Case Western Reserve University.

Appellate Advocacy, 1984, approximately 270 pages, by Alan D. Hornstein, Professor of Law, University of Maryland.

Art Law, 1984, approximately 290 pages, by Leonard D. DuBoff, Professor of Law, Lewis and Clark College, Northwestern School of Law.

Banking and Financial Institutions, 1984, 409 pages, by William A. Lovett, Professor of Law, Tulane University.

Church-State Relations—Law of, 1981, 305 pages, by Leonard F. Manning, Late Professor of Law, Fordham University.

I

NUTSHELL SERIES

Civil Procedure, 1979, 271 pages, by Mary Kay Kane, Professor of Law, University of California, Hastings College of the Law.

Civil Rights, 1978, 279 pages, by Norman Vieira, Professor of Law, Southern Illinois University.

Commercial Paper, 3rd Ed., 1982, 404 pages, by Charles M. Weber, Professor of Business Law, University of Arizona and Richard E. Speidel, Professor of Law, Northwestern University.

Community Property, 1982, 423 pages, by Robert L. Mennell, Professor of Law, Hamline University.

Comparative Legal Traditions, 1982, 402 pages, by Mary Ann Glendon, Professor of Law, Boston College, Michael Wallace Gordon, Professor of Law, University of Florida and Christopher Osakwe, Professor of Law, Tulane University.

Conflicts, 1982, 469 pages, by David D. Siegel, Professor of Law, Albany Law School, Union University.

Constitutional Analysis, 1979, 388 pages, by Jerre S. Williams, Professor of Law Emeritus, University of Texas.

Constitutional Power—Federal and State, 1974, 411 pages, by David E. Engdahl, Professor of Law, University of Puget Sound.

Consumer Law, 2nd Ed., 1981, 418 pages, by David G. Epstein, Professor of Law, University of Texas and Steve H. Nickles, Professor of Law, University of Minnesota.

Contract Remedies, 1981, 323 pages, by Jane M. Friedman, Professor of Law, Wayne State University.

Contracts, 2nd Ed., 1983, approximately 410 pages, by Gordon D. Schaber, Dean and Professor of Law, McGeorge School of Law and Claude D. Rohwer, Professor of Law, McGeorge School of Law.

Corporations—Law of, 1980, 379 pages, by Robert W. Hamilton, Professor of Law, University of Texas.

Corrections and Prisoners' Rights—Law of, 2nd Ed., 1983, 384 pages, by Sheldon Krantz, Dean and Professor of Law, University of San Diego.

NUTSHELL SERIES

Criminal Law, 1975, 302 pages, by Arnold H. Loewy, Professor of Law, University of North Carolina.

Criminal Procedure—Constitutional Limitations, 3rd Ed., 1980, 438 pages, by Jerold H. Israel, Professor of Law, University of Michigan and Wayne R. LaFave, Professor of Law, University of Illinois.

Debtor-Creditor Law, 2nd Ed., 1980, 324 pages, by David G. Epstein, Professor of Law, University of Texas.

Employment Discrimination—Federal Law of, 2nd Ed., 1981, 402 pages, by Mack A. Player, Professor of Law, University of Georgia.

Energy Law, 1981, 338 pages, by Joseph P. Tomain, Professor of Law, Drake University.

Environmental Law, 1983, 343 pages by Roger W. Findley, Professor of Law, University of Illinois and Daniel A. Farber, Professor of Law, University of Minnesota.

Estate Planning—Introduction to, 3rd Ed., 1983, 370 pages, by Robert J. Lynn, Professor of Law, Ohio State University.

Evidence, Federal Rules of, 1981, 428 pages, by Michael H. Graham, Professor of Law, University of Illinois.

Evidence, State and Federal Rules, 2nd Ed., 1981, 514 pages, by Paul F. Rothstein, Professor of Law, Georgetown University.

Family Law, 1977, 400 pages, by Harry D. Krause, Professor of Law, University of Illinois.

Federal Estate and Gift Taxation, 3rd Ed., 1983, 509 pages, by John K. McNulty, Professor of Law, University of California, Berkeley.

Federal Income Taxation of Individuals, 3rd Ed., 1983, 487 pages, by John K. McNulty, Professor of Law, University of California, Berkeley.

Federal Income Taxation of Corporations and Stockholders, 2nd Ed., 1981, 362 pages, by Jonathan Sobeloff, Late Professor of Law, Georgetown University and Peter P. Weidenbruch, Jr., Professor of Law, Georgetown University.

Federal Jurisdiction, 2nd Ed., 1981, 258 pages, by David P. Currie, Professor of Law, University of Chicago.

NUTSHELL SERIES

Future Interests, 1981, 361 pages, by Lawrence W. Waggoner, Professor of Law, University of Michigan.

Government Contracts, 1979, 423 pages, by W. Noel Keyes, Professor of Law, Pepperdine University.

Historical Introduction to Anglo-American Law, 2nd Ed., 1973, 280 pages, by Frederick G. Kempin, Jr., Professor of Business Law, Wharton School of Finance and Commerce, University of Pennsylvania.

Immigration Law and Procedure, 1984, approximately 240 pages, by David Weissbrodt, Professor of Law, University of Minnesota.

Injunctions, 1974, 264 pages, by John F. Dobbyn, Professor of Law, Villanova University.

Insurance Law, 1981, 281 pages, by John F. Dobbyn, Professor of Law, Villanova University.

Intellectual Property—Patents, Trademarks and Copyright, 1983, 428 pages, by Arthur R. Miller, Professor of Law, Harvard University, and Michael H. Davis, Professor of Law, Cleveland State University, Cleveland-Marshall College of Law.

International Business Transactions, 2nd Ed., 1984, 476 pages, by Donald T. Wilson, Professor of Law, Loyola University, Los Angeles.

Introduction to the Study and Practice of Law, 1983, 418 pages, by Kenney F. Hegland, Professor of Law, University of Arizona.

Judicial Process, 1980, 292 pages, by William L. Reynolds, Professor of Law, University of Maryland.

Jurisdiction, 4th Ed., 1980, 232 pages, by Albert A. Ehrenzweig, Late Professor of Law, University of California, Berkeley, David W. Louisell, Late Professor of Law, University of California, Berkeley and Geoffrey C. Hazard, Jr., Professor of Law, Yale Law School.

Juvenile Courts, 3rd Ed., 1984, approximately 260 pages, by Sanford J. Fox, Professor of Law, Boston College.

NUTSHELL SERIES

Labor Arbitration Law and Practice, 1979, 358 pages, by Dennis R. Nolan, Professor of Law, University of South Carolina.

Labor Law, 1979, 403 pages, by Douglas L. Leslie, Professor of Law, University of Virginia.

Land Use, 1978, 316 pages, by Robert R. Wright, Professor of Law, University of Arkansas, Little Rock and Susan Webber, Professor of Law, University of Arkansas, Little Rock.

Landlord and Tenant Law, 1979, 319 pages, by David S. Hill, Professor of Law, University of Colorado.

Law Study and Law Examinations—Introduction to, 1971, 389 pages, by Stanley V. Kinyon, Late Professor of Law, University of Minnesota.

Legal Interviewing and Counseling, 1976, 353 pages, by Thomas L. Shaffer, Professor of Law, Washington and Lee University.

Legal Research, 3rd Ed., 1978, 415 pages, by Morris L. Cohen, Professor of Law and Law Librarian, Yale University.

Legal Writing, 1982, 294 pages, by Dr. Lynn B. Squires and Marjorie Dick Rombauer, Professor of Law, University of Washington.

Legislative Law and Process, 1975, 279 pages, by Jack Davies, Professor of Law, William Mitchell College of Law.

Local Government Law, 2nd Ed., 1983, 404 pages, by David J. McCarthy, Jr., Professor of Law, Georgetown University.

Mass Communications Law, 2nd Ed., 1983, 473 pages, by Harvey L. Zuckman, Professor of Law, Catholic University and Martin J. Gaynes, Lecturer in Law, Temple University.

Medical Malpractice—The Law of, 1977, 340 pages, by Joseph H. King, Professor of Law, University of Tennessee.

Military Law, 1980, 378 pages, by Charles A. Shanor, Professor of Law, Emory University and Timothy P. Terrell, Professor of Law, Emory University.

Oil and Gas, 1983, 443 pages, by John S. Lowe, Professor of Law, University of Tulsa.

Personal Property, 1983, 322 pages, by Barlow Burke, Jr., Professor of Law, American University.

NUTSHELL SERIES

Post-Conviction Remedies, 1978, 360 pages, by Robert Popper, Professor of Law, University of Missouri, Kansas City.

Presidential Power, 1977, 328 pages, by Arthur Selwyn Miller, Professor of Law Emeritus, George Washington University.

Procedure Before Trial, 1972, 258 pages, by Delmar Karlen, Professor of Law, College of William and Mary.

Products Liability, 2nd Ed., 1981, 341 pages, by Dix W. Noel, Late Professor of Law, University of Tennessee and Jerry J. Phillips, Professor of Law, University of Tennessee.

Professional Responsibility, 1980, 399 pages, by Robert H. Aronson, Professor of Law, University of Washington, and Donald T. Weckstein, Professor of Law, University of San Diego.

Real Estate Finance, 1979, 292 pages, by Jon W. Bruce, Professor of Law, Vanderbilt University.

Real Property, 2nd Ed., 1981, 448 pages, by Roger H. Bernhardt, Professor of Law, Golden Gate University.

Regulated Industries, 1982, 394 pages, by Ernest Gellhorn, Dean and Professor of Law, Case Western Reserve University, and Richard J. Pierce, Professor of Law, Tulane University.

Remedies, 1977, 364 pages, by John F. O'Connell, Professor of Law, Western State University College of Law, Fullerton.

Res Judicata, 1976, 310 pages, by Robert C. Casad, Professor of Law, University of Kansas.

Sales, 2nd Ed., 1981, 370 pages, by John M. Stockton, Professor of Business Law, Wharton School of Finance and Commerce, University of Pennsylvania.

Schools, Students and Teachers—Law of, 1984, approximately 350 pages, by Kern Alexander, Professor of Education, University of Florida and M. David Alexander, Professor, Virginia Tech University.

Sea—Law of, 1984, approximately 250 pages, by Louis B. Sohn, Professor of Law, Harvard University and Kristen Gustafson.

Secured Transactions, 2nd Ed., 1981, 391 pages, by Henry J. Bailey, Professor of Law Emeritus, Willamette University.

NUTSHELL SERIES

Securities Regulation, 2nd Ed., 1982, 322 pages, by David L. Ratner, Dean and Professor of Law, University of San Francisco.

Sex Discrimination, 1982, 399 pages, by Claire Sherman Thomas, Lecturer, University of Washington, Women's Studies Department.

Titles—The Calculus of Interests, 1968, 277 pages, by Oval A. Phipps, Late Professor of Law, St. Louis University.

Torts—Injuries to Persons and Property, 1977, 434 pages, by Edward J. Kionka, Professor of Law, Southern Illinois University.

Torts—Injuries to Family, Social and Trade Relations, 1979, 358 pages, by Wex S. Malone, Professor of Law Emeritus, Louisiana State University.

Trial Advocacy, 1979, 402 pages, by Paul B. Bergman, Adjunct Professor of Law, University of California, Los Angeles.

Trial and Practice Skills, 1978, 346 pages, by Kenney F. Hegland, Professor of Law, University of Arizona.

Trial, The First—Where Do I Sit? What Do I Say?, 1982, 396 pages, by Steven H. Goldberg, Professor of Law, University of Minnesota.

Unfair Trade Practices, 1982, 444 pages, by Charles R. McManis, Professor of Law, Washington University, St. Louis.

Uniform Commercial Code, 1975, 507 pages, by Bradford Stone, Professor of Law, Detroit College of Law.

Uniform Probate Code, 1978, 425 pages, by Lawrence H. Averill, Jr., Dean and Professor of Law, University of Arkansas, Little Rock.

Water Law, 1984, approximately 400 pages, by David H. Getches, Professor of Law, University of Colorado.

Welfare Law—Structure and Entitlement, 1979, 455 pages, by Arthur B. LaFrance, Dean and Professor of Law, Lewis and Clark College, Northwestern School of Law.

Wills and Trusts, 1979, 392 pages, by Robert L. Mennell, Professor of Law, Hamline University.

NUTSHELL SERIES

Workers' Compensation and Employee Protection Laws, 1984, 248 pages, by Jack B. Hood, Professor of Law, Cumberland School of Law, Samford University and Benjamin A. Hardy, Professor of Law, Cumberland School of Law, Samford University.

Hornbook Series
and
Basic Legal Texts
of
WEST PUBLISHING COMPANY
P.O. Box 43526
St. Paul, Minnesota 55164
January, 1984

Administrative Law, Davis' Text on, 3rd Ed., 1972, 617 pages, by Kenneth Culp Davis, Professor of Law, University of San Diego.

Agency, Seavey's Hornbook on, 1964, 329 pages, by Warren A. Seavey, Late Professor of Law, Harvard University.

Agency and Partnership, Reuschlein & Gregory's Hornbook on the Law of, 1979 with 1981 Pocket Part, 625 pages, by Harold Gill Reuschlein, Professor of Law, St. Mary's University and William A. Gregory, Professor of Law, Georgia State University.

Antitrust, Sullivan's Hornbook on the Law of, 1977, 886 pages, by Lawrence A. Sullivan, Professor of Law, University of California, Berkeley.

Common Law Pleading, Koffler and Reppy's Hornbook on, 1969, 663 pages, by Joseph H. Koffler, Professor of Law, New York Law School and Alison Reppy, Late Dean and Professor of Law, New York Law School.

Common Law Pleading, Shipman's Hornbook on, 3rd Ed., 1923, 644 pages, by Henry W. Ballentine, Late Professor of Law, University of California, Berkeley.

Conflict of Laws, Scoles and Hay's Hornbook on, Student Ed., 1982, 1085 pages, by Eugene F. Scoles, Professor of Law,

HORNBOOKS & BASIC TEXTS

University of Illinois and Peter Hay, Dean and Professor of Law, University of Illinois.

Constitutional Law, Nowak, Rotunda and Young's Hornbook on, 2nd Ed., Student Ed., 1983, 1172 pages, by John E. Nowak, Professor of Law, University of Illinois, Ronald D. Rotunda, Professor of Law, University of Illinois, and J. Nelson Young, Professor of Law, University of North Carolina.

Contracts, Calamari and Perillo's Hornbook on, 2nd Ed., 1977, 878 pages, by John D. Calamari, Professor of Law, Fordham University and Joseph M. Perillo, Professor of Law, Fordham University.

Contracts, Corbin's One Volume Student Ed., 1952, 1224 pages, by Arthur L. Corbin, Late Professor of Law, Yale University.

Contracts, Simpson's Hornbook on, 2nd Ed., 1965, 510 pages, by Laurence P. Simpson, Late Professor of Law, New York University.

Corporate Taxation, Kahn's Handbook on, 3rd Ed., Student Ed., Soft cover, 1981 with 1983 Supplement, 614 pages, by Douglas A. Kahn, Professor of Law, University of Michigan.

Corporations, Henn and Alexander's Hornbook on, 3rd Ed., Student Ed., 1983, 1371 pages, by Harry G. Henn, Professor of Law, Cornell University and John R. Alexander, Member, New York and Hawaii Bars.

Criminal Law, LaFave and Scott's Hornbook on, 1972, 763 pages, by Wayne R. LaFave, Professor of Law, University of Illinois, and Austin Scott, Jr., Late Professor of Law, University of Colorado.

Damages, McCormick's Hornbook on, 1935, 811 pages, by Charles T. McCormick, Late Dean and Professor of Law, University of Texas.

Domestic Relations, Clark's Hornbook on, 1968, 754 pages, by Homer H. Clark, Jr., Professor of Law, University of Colorado.

Environmental Law, Rodgers' Hornbook on, 1977, 956 pages, by William H. Rodgers, Jr., Professor of Law, University of Washington.

HORNBOOKS & BASIC TEXTS

Evidence, Lilly's Introduction to, 1978, 486 pages, by Graham C. Lilly, Professor of Law, University of Virginia.

Evidence, McCormick's Hornbook on, 3rd Ed., Student Ed., 1984, approximately 1028 pages, General Editor, Edward W. Cleary, Professor of Law Emeritus, Arizona State University.

Federal Courts, Wright's Hornbook on, 4th Ed., Student Ed., 1983, 870 pages, by Charles Alan Wright, Professor of Law, University of Texas.

Federal Income Taxation of Individuals, Posin's Hornbook on, Student Ed., 1983, 491 pages, by Daniel Q. Posin, Jr., Professor of Law, Hofstra University.

Future Interest, Simes' Hornbook on, 2nd Ed., 1966, 355 pages, by Lewis M. Simes, Late Professor of Law, University of Michigan.

Insurance, Keeton's Basic Text on, 1971, 712 pages, by Robert E. Keeton, Professor of Law Emeritus, Harvard University.

Labor Law, Gorman's Basic Text on, 1976, 914 pages, by Robert A. Gorman, Professor of Law, University of Pennsylvania.

Law Problems, Ballentine's, 5th Ed., 1975, 767 pages, General Editor, William E. Burby, Late Professor of Law, University of Southern California.

Legal Writing Style, Weihofen's, 2nd Ed., 1980, 332 pages, by Henry Weihofen, Professor of Law Emeritus, University of New Mexico.

Local Government Law, Reynolds' Hornbook on, 1982, 860 pages, by Osborne M. Reynolds, Professor of Law, University of Oklahoma.

New York Practice, Siegel's Hornbook on, 1978, with 1981–82 Pocket Part, 1011 pages, by David D. Siegel, Professor of Law, Albany Law School of Union University.

Oil and Gas, Hemingway's Hornbook on, 2nd Ed., Student Ed., 1983, 543 pages, by Richard W. Hemingway, Professor of Law, University of Oklahoma.

Poor, Law of the, LaFrance, Schroeder, Bennett and Boyd's Hornbook on, 1973, 558 pages, by Arthur B. LaFrance, Dean

HORNBOOKS & BASIC TEXTS

and Professor of Law, Lewis and Clark College, Northwestern School of Law, Milton R. Schroeder, Professor of Law, Arizona State University, Robert W. Bennett, Professor of Law, Northwestern University and William E. Boyd, Professor of Law, University of Arizona.

Property, Boyer's Survey of, 3rd Ed., 1981, 766 pages, by Ralph E. Boyer, Professor of Law, University of Miami.

Property, Law of, Cunningham, Whitman and Stoebuck's Hornbook on, Student Ed., 1984, approximately 808 pages, by Roger A. Cunningham, Professor of Law, University of Michigan, Dale A. Whitman, Dean and Professor of Law, University of Missouri–Columbia and William B. Stoebuck, Professor of Law, University of Washington.

Real Estate Finance Law, Osborne, Nelson and Whitman's Hornbook on, (successor to Hornbook on Mortgages), 1979, 885 pages, by George E. Osborne, Late Professor of Law, Stanford University, Grant S. Nelson, Professor of Law, University of Missouri, Columbia and Dale A. Whitman, Dean and Professor of Law, University of Missouri, Columbia.

Real Property, Burby's Hornbook on, 3rd Ed., 1965, 490 pages, by William E. Burby, Late Professor of Law, University of Southern California.

Real Property, Moynihan's Introduction to, 1962, 254 pages, by Cornelius J. Moynihan, Professor of Law, Suffolk University.

Remedies, Dobb's Hornbook on, 1973, 1067 pages, by Dan B. Dobbs, Professor of Law, University of Arizona.

Sales, Nordstrom's Hornbook on, 1970, 600 pages, by Robert J. Nordstrom, former Professor of Law, Ohio State University.

Secured Transactions under the U.C.C., Henson's Hornbook on, 2nd Ed., 1979, with 1979 Pocket Part, 504 pages, by Ray D. Henson, Professor of Law, University of California, Hastings College of the Law.

Torts, Prosser and Keeton's Hornbook on, 5th Ed., Student Ed., 1984, approximately 1052 pages, by William L. Prosser, Late Dean and Professor of Law, University of California, Berkeley, Page Keeton, Professor of Law, University of Texas, Dan B. Dobbs, Professor of Law University of Arizona, Rob-

HORNBOOKS & BASIC TEXTS

ert E. Keeton, Professor of Law Emeritus, Harvard University and David G. Owen, Professor of Law, University of South Carolina.

Trial Advocacy, Jeans' Handbook on, Student Ed., Soft cover, 1975, by James W. Jeans, Professor of Law, University of Missouri, Kansas City.

Trusts, Bogert's Hornbook on, 5th Ed., 1973, 726 pages, by George G. Bogert, Late Professor of Law, University of Chicago and George T. Bogert, Attorney, Chicago, Illinois.

Urban Planning and Land Development Control, Hagman's Hornbook on, 1971, 706 pages, by Donald G. Hagman, Late Professor of Law, University of California, Los Angeles.

Uniform Commercial Code, White and Summers' Hornbook on, 2nd Ed., 1980, 1250 pages, by James J. White, Professor of Law, University of Michigan and Robert S. Summers, Professor of Law, Cornell University.

Wills, Atkinson's Hornbook on, 2nd Ed., 1953, 975 pages, by Thomas E. Atkinson, Late Professor of Law, New York University.

Advisory Board

Professor JOHN A. BAUMAN
University of California School of Law, Los Angeles

Professor CURTIS J. BERGER
Columbia University School of Law

Dean JESSE H. CHOPER
University of California School of Law, Berkeley

Professor DAVID P. CURRIE
University of Chicago Law School

Professor DAVID G. EPSTEIN
University of Texas School of Law

Dean ERNEST GELLHORN
Case Western Reserve University,
Franklin T. Backus Law School

Professor YALE KAMISAR
University of Michigan Law School

Professor WAYNE R. LaFAVE
University of Illinois College of Law

Professor RICHARD C. MAXWELL
Duke University School of Law

Professor ARTHUR R. MILLER
Harvard University Law School

Professor JAMES J. WHITE
University of Michigan Law School

Professor CHARLES ALAN WRIGHT
University of Texas School of Law

COMMERCIAL PAPER
IN A NUTSHELL

By

CHARLES M. WEBER
Professor of Business Law Emeritus
at the Wharton School of the
University of Pennsylvania

and

RICHARD E. SPEIDEL
Professor of Law, Northwestern University
School of Law

Third Edition

ST. PAUL, MINN.
WEST PUBLISHING CO.
1982

COPYRIGHT © 1982 by WEST PUBLISHING CO.
50 West Kellogg Boulevard
P. O. Box 3526
St. Paul, Minnesota 55165

All rights reserved
Printed in the United States of America

Library of Congress Cataloging in Publication Data

Weber, Charles M., 1911–
  Commercial paper in a nutshell.

  (Nutshell series)
  Includes index.
  1. Negotiable instruments—United States. I.
Speidel, Richard E. II. Title. III. Series.
KF957.Z9W4 1982   346.73'096   82–2027
                          347.30696   AACR2

ISBN 0–314–65059–8

Weber & Speidel, Comm.Paper, 3rd Ed.
  1st Reprint—1984

To

Ida Elizabeth

and

Dorothy Jean

PREFACE

The purpose of this book is to introduce the law of commercial paper in a way that promotes understanding and furnishes a sound foundation for any further study of the subject that the reader may pursue.

As used in the title and throughout the book, "commercial paper" refers to promissory notes, drafts, checks and certificates of deposit—the subject matter of Article 3 and, when checks are involved, Article 4 of the Uniform Commercial Code. It does not purport to treat other types of documents that are often used in commercial transactions, such as bills of lading, warehouse receipts, security agreements, letters of credit and investment securities, all of which are treated in other articles of the Code. The book does treat the scope of Articles 3 and 4, however, and their relationship to other relevant Code articles.

In the spirit of the NUTSHELL, the emphasis is on basic principles. We focus upon recurring transactions and the Code sections applicable to help structure and to resolve disputes arising under them. In some cases, the relevant sections of Articles 3 and 4 are quoted in full. In others, they are simply paraphrased or cited. In all cases, the expectation is that the reader will supplement this text with a copy of the Uniform Commercial Code and read the sections cited. Text and Code must go hand in hand to achieve the purpose of this book.

Occasionally, statements in the text are made in reliance on Official Comments to the statute. It is true that

PREFACE

the comments have not been enacted into law and that some of them have been questioned. Nonetheless, they were prepared and sponsored by the American Law Institute and the Conference of Commissioners on Uniform State Laws and are often a source of help when one is uncertain about the meaning or rationale of a Code provision. Less frequently, reference is made to recent cases or secondary material dealing with the questions under discussion. In the last few years, the number of decided cases and law review articles has multiplied and they are beyond the capacity of this book fully to discuss or cite. They should, of course, be consulted when one seeks to broaden the scope and depth of the inquiry.

The Third Edition hews closely to the organization and structure of the Second. Major changes include a rewriting of Chapter 1, a complete revision and elaboration of Chapter 2 (now Chapter 3), the addition of material on check collections under Article 4 in Chapter 13 and a new Chapter 14 on electronic fund transfers. Similarly, the Third Edition develops more thoroughly the scope of Articles 3 and 4 and their relationship to other Code articles (see Chapter 1, Section 5, Chapter 2, Section 3 and Chapter 3), updates the status of the holder-in-due-course in consumer credit transactions (see Chapter 10, Section 14) and adds several new sections on such subjects as "accrual of cause of action," (Chapter 7, Section 5) and "collecting bank as holder in due course" (see Chapter 10, Section 13). Finally, the text has been modified, where appropriate, to reflect recent developments, and the short, illustrative cases in the Second Edition have been greatly expanded in number. There are 84 "examples" in the Third Edition, which are designed both to illustrate basic principles and to reflect case law development since 1975.

PREFACE

As always, there are many people who should be thanked when a book is completed. My thanks go to Professor Charles M. Weber, author of the First and Second Editions. As NUTSHELLS go, his was excellent. I have simply tried to preserve the tradition.

RICHARD E. SPEIDEL

March, 1982

*

OUTLINE

	Page
Preface	XIX
Table of Cases	XXXIX
Table of References to U.C.C. Provisions and Comments	XLI

CHAPTER 1. INTRODUCTION 1
Section
1. Importance of Commercial Paper 1
2. Early Law of Commercial Paper 3
3. Uniform Negotiable Instruments Law 5
4. Uniform Commercial Code 5
5. The Scope of UCC Article 3 8
   A. Instruments are Within Scope 8
   B. Exclusions from Scope 10
      Example 1 11
      Example 2 13

CHAPTER 2. FORMS AND USES OF COMMERCIAL PAPER AND TERMINOLOGY 14
Section
1. Forms of Commercial Paper and Parties 14
   A. Promissory Note 14
   B. Draft 17
   C. Check 18
   D. Certificate of Deposit 21
2. Uses of Commercial Paper 23
   A. Promissory Notes 23
      1. Means of Borrowing Money 23
      2. Means of Buying on Credit 25

OUTLINE

Section

2. Uses of Commercial Paper—Continued
    3. Means of Evidencing Previously Existing Debt 26
    B. Draft 26
       1. Means of Collecting Accounts 26
       2. Means of Financing the Movement of Goods 28
    C. Check 31
    D. Certificate of Deposit 32
3. Functional Overlaps Among Articles of Code 34

CHAPTER 3. NEGOTIABILITY AND RISK ALLOCATION IN COMMERCIAL PAPER 36

Section

1. Assignment of Contracts Rights 37
    A. Development at Common Law 37
       Example 3 40
    B. Effect of UCC Article 9, Secured Transactions 41
       Example 4 43
2. Negotiability Under Article 3: A Contrast with Assignment 43
       Example 5 47
3. Why Negotiability? 47

CHAPTER 4. REQUISITES OF NEGOTIABILITY 49

Section

1. Requisites of Negotiability Under Article 3: 3-104(1) 49
    A. Must Be in Writing: 3-104(1) 51
    B. Must Be Signed by Maker or Drawer: 3-104(1)(a) 52

XXIV

OUTLINE

1. Requisites of Negotiability Under Article 3: 3-104(1)—Continued
   - C. Must Contain a Promise or Order: 3-104(1)(b) 52
     1. Promise 52
     2. Order 53
        a. Drawer Named as Drawee 54
        b. Multiple Drawees 55
   - D. Promise or Order Must Be Unconditional: 3-104(1)(b); 3-105 55
     1. Terms of Instrument Determine Presence of Condition 56
     2. Effect of Another Writing: 3-119 ... 57
     3. Implied Conditions Not Recognized .. 58
     4. "Subject to" or Similar Words Negate Negotiability 59
     5. Reference to Account or Fund 60
     Example 6 61
   - E. Must Specify a Sum Certain: 3-104(1)(b); 3-106 62
   - F. Must be Payable in Money: 3-104(1)(b); 3-107 64
     Example 7 64
   - G. Must Be Payable on Demand or at a Definite Time: 3-104(1)(c); 3-108; 3-109 65
     1. Payable on Demand 65
     2. Payable at a Definite Time 66
     3. Acceleration Clauses 66
     4. Extension Clauses 67
     5. Event Certain to Happen 68
     Example 8 68
   - H. Must Be Payable to Order or to Bearer: 3-104(1)(d); 3-110; 3-111 69
     1. Payable to Order 70
     2. Payable to Bearer 72

OUTLINE

Section

1. Requisites of Negotiability Under Article 3: 3–104(1)—Continued **Page**

    I. Must Contain no Other Promise, Order, Obligation or Power 73
2. Instrument Declared to Be Negotiable 74
3. Instrument Lacking Only Words of Negotiability 75

    Example 9 76
4. Terms and Omissions Not Affecting Negotiability: 3–112 77

    Example 10 79

    Example 11 79
5. Incomplete Instruments 80
6. Omitting Date, Antedating, and Postdating .. 80

    Example 12 81
7. "Instrument," "Note," "Draft," and "Check" Presumed to Refer to Instrument That Is Negotiable 82

CHAPTER 5. ISSUE: RELATIONSHIP BETWEEN THE IMMEDIATE PARTIES TO AN INSTRUMENT 83

Section

1. Meaning of "Issue" 83
2. Meaning of "Delivery" 85
3. Effect of Delivery—Contract Between Immediate Parties 86
4. Interpretation and Scope of the Agreement .. 87

    A. Interpretation 87

    Example 13 88

    B. Scope: The Parol Evidence Rule 88

    1. Limitations on Operation of Parol Evidence Rule 91

OUTLINE

Section
4. Interpretation and Scope of the Agreement
—Continued

		Page
2.	Conditional Delivery and the Parol Evidence Rule	93
3.	Understanding Signer Is Not to Be Bound	95

5. Similar Principles Apply to Subsequent Delivery 96
   Example 14 97

CHAPTER 6. TRANSFER AND NEGOTIATION 99

Section
1. Negotiation and Assignment 99
2. Requisites of Negotiation 100
3. Bearer and Order Paper Distinguished 101
4. Blank and Special Indorsements Distinguished 101
5. Appropriate Party to Indorse 102
   A. Holder 103
6. Requisites of Negotiation Illustrated 104
   Example 15 104
   Example 16 104
   Example 17 105
   Example 18 105
   Example 19 105
7. "Holder" Not Synonymous with "Owner" 106
8. Risks of Bearer and Order Paper 106
9. Fictitious Payee Problem 108
   Example 20 111
10. Imposter Problem 113
11. Incomplete Negotiation 116
    Example 21 117
12. Depositary Bank's Power to Supply Missing Indorsement to Complete Negotiation 118
13. Negotiation by Multiple Payees 119

XXVII

OUTLINE

Section		Page
14.	Effect of Attempt to Negotiate Less Than Balance Due	120
15.	Place of Indorsement on Instrument	120
16.	Negotiation When Name Is Misspelled or Misstated	121
17.	Effect of Negotiation That May Be Rescinded	121
18.	Unqualified and Qualified Indorsements Affect Liability	122
19.	Restrictive Indorsements	123
	A. Indorsements That Purport to Prohibit Further Transfer	124
	B. Conditional Indorsement	124
	C. Indorsement for Deposit or Collection	125
	1. Depositary Banks and Transferees Outside Banking Process	126
	2. Special Treatment of Intermediary and Payor Banks	129
	D. Trust Indorsements	133
	Example 22	134
	E. Effect of Restrictive Indorsement on Discharge of Payor	135
	Example 23	135
20.	Words of Assignment, Waiver, Guarantee, Limitation or Disclaimer	136

CHAPTER 7. LIABILITY OF PARTIES ON THE INSTRUMENT 137

Section		
1.	Various Bases of Liability	137
2.	General Principles Governing Liability on the Instrument	138
	A. Person's Signature Must Appear on Instrument	138
	B. Holder's Name Need Not Appear to Give Right to Sue on Instrument	139
	C. Signature	140

XXVIII

OUTLINE

Section

2. General Principles Governing Liability on the Instrument—Continued
   D. Determining Capacity in Which Party Signs 140
   E. Signature by Authorized Representative 141
      Example 24 145
   F. Unauthorized Signatures 145
   G. Alteration 146
      Example 25 148
   H. Negligence Contributing to Unauthorized Signature or Alteration 149
3. Liability of Particular Parties on the Instrument 151
   A. In General 151
   B. Maker's Liability 152
   C. Drawer's Liability 152
   D. Drawee's Liability 153
      Example 26 154
   E. Acceptor's Liability 155
      Example 27 157
      1. Acceptance Varying Draft 158
   F. Unqualified Indorser's Secondary Liability 159
      Example 28 161
   G. Qualified Indorser's Liability 161
      Example 29 162
   H. Liability for Conversion 163
      Example 30 163
4. Special Kinds of Liability on the Instrument 163
   A. Accommodation and Accommodated Parties 164
      Example 31 167
   B. Words of Guarantee in an Indorsement .. 167
      Example 32 168
5. Accrual of Cause of Action 169
   A. Against Maker or Acceptor 170
   B. Against Drawer and Indorser 171

XXIX

OUTLINE

Section

5. Accrual of Cause of Action—Continued **Page**
   C. Against Guarantor 172
   D. Against Accommodation Party 173
   E. Interest 173

CHAPTER 8. PRESENTMENT, DISHONOR, NOTICE OF DISHONOR, AND PROTEST 175

Section

1. Effect of Failing to Satisfy Conditions 176
   A. Indorsers 176
      1. Reason for Discharging Secondary Liability of Indorser—Recoupment ... 177
   B. Drawers 179
      Example 33 181
      Example 34 182
   C. Drawees 183
   D. Makers and Acceptors 183
2. Satisfying Requirements 184
   A. Presentment 185
      1. What Constitutes Effective Presentment 185
      Example 35 187
      2. Time of Presentment 188
         a. Presentment for Acceptance 188
         b. Presentment for Payment 189
         c. Reasonable Time for Presentment 191
         d. Presentment by Collecting Bank 193
         e. Domiciled Instrument 193
   B. Dishonor 195
      1. What Constitutes 195
   C. Notice of Dishonor 197
      1. Method of Giving Notice 197
      2. Persons Given Notice 198
      3. Persons Giving Notice 199

OUTLINE

Section
2. Satisfying Requirements—Continued   **Page**
      4. Sequence in Which Notice May Be Given 199
      5. Time Allowed for Giving Notice 199
  D. Protest 200
      1. When Protest Is Required 201
      2. Time for Making Protest 201
      3. Protest as Evidence 202
      4. Forwarding Protest 202
      5. Other Types of Evidence 203
3. Excusing Omission or Delay in Presentment, Notice or Protest 203
4. Effect of Excusing Omission or Delay 205
      Example 36 205

CHAPTER 9. LIABILITY BASED ON WARRANTIES 208

Section
1. Transferors' Warranties 208
  A. Why Holder May Prefer to Sue Transferor on Warranty 209
  B. Warranty of Title 210
      Example 37 212
  C. Warranty Signatures are Genuine or Authorized 212
      Example 38 213
      Example 39 214
  D. Warranty Against Material Alteration ... 214
      Example 40 215
  E. Warranty Regarding Defenses 215
      Example 41 216
  F. Warranty of No Knowledge of Insolvency Proceedings 216
      Example 42 217
  G. Holder in Due Course and Transferor's Warranties 217

OUTLINE

Section		Page
2.	Presenters' Warranties	218
	A. The Doctrine of Finality	220
	Example 43	221
	B. Warranty of Title	222
	Example 44	222
	C. Warranty Against Knowledge Signature of Maker or Drawer Is Unauthorized	225
	Example 45	225
	Example 46	228
	Example 47	228
	D. Warranty Against Material Alterations	229
	E. Presenters' Warranties Given Also by Prior Transferors	232
	Example 48	232
	F. Transferees of Holder in Due Course	233
	Example 49	233
3.	Remedies Available for Breach of Warranty	233
4.	Warranties in Check Collection Process	235

CHAPTER 10. HOLDERS IN DUE COURSE — 237

Section		
1.	Importance	237
2.	Requirements	238
3.	Instrument	238
4.	Holder	239
5.	Value	239
	A. Executory Promise Normally Not Value	241
	Example 50	242
	B. Notice of Claim or Defense While Promise Is Executory	242
	Example 51	243
	C. Executory Promise as Value	243
	D. Bank Credit as Value	245
	Example 52	246
	E. Taking Instrument as Payment or Security for Antecedent Debt as Value	247
	Example 53	248

OUTLINE

Section		Page
6.	Good Faith	249
	Example 54	251
7.	Without Notice	251
	A. Notice Instrument Is Overdue	254
	B. Notice Instrument Has Been Dishonored	257
	C. Notice of Defense Against Instrument or Claim to It	258
	1. Discharge	260
	Example 55	260
	2. Counterclaims and Set-Offs	260
	Example 56	261
	3. Notice Promise of Prior Party Is Unperformed	261
	4. Instrument as Notice of Defense or Claim	261
	5. Fiduciaries	263
8.	Notice of One Defense or Claim Subjects Holder to All	264
	Example 57	264
9.	Payee as Holder in Due Course	264
	Example 58	265
10.	Taking Through Holder in Due Course—The Shelter Provision	266
	Example 59	266
	Example 60	267
11.	Transactions that Cannot Give Rise to Holder in Due Course	267
12.	Assignee with Principal Advantage of Holder in Due Course	268
13.	Collecting Bank as Holder in Due Course	269
14.	Special Treatment of Consumers	270
	A. The Courts	272
	1. Lack of Good Faith	273

OUTLINE

Section

14. Special Treatment of Consumers—Continued
    2. Unity of Dealer and Financial Institution _____ 274
    3. Supporting Arguments _____ 275
    B. The Legislatures: Uniform Consumer Credit Code _____ 276
    C. Federal Trade Commission _____ 279
       Example 61 _____ 280

CHAPTER 11. DEFENSES AND CLAIMS ____ 283

Section

1. Defenses Effective Against Holder in Due Course _____ 283
    A. Defenses Arising at Time of Transfer or Later _____ 283
    B. Defenses Arising Before Negotiation to Holder in Due Course _____ 284
        1. Infancy _____ 285
        2. Other Incapacity _____ 286
        3. Duress _____ 287
        4. Illegality _____ 287
        5. Fraud _____ 288
        Example 62 _____ 289
        6. Discharge in Insolvency Proceedings 290
        7. Discharge of Which Holder in Due Course Has Notice _____ 291
        8. Material Alteration _____ 291
        9. Unauthorized Signature Including Forgery _____ 292
2. Defenses Against One who Lacks the Rights of a Holder in Due Course _____ 293
    A. Outside Agreement _____ 295
    B. Defense That Third Party Has Claim—Jus Tertii _____ 297
       Example 63 _____ 297

XXXIV

OUTLINE

Section		Page
3.	Claims Asserted Against Holder in Due Course	298
4.	Claims Asserted Against Person Lacking Rights of Holder in Due Course	299
5.	Procedure	299
	A. Plaintiff	300
	B. Burden of Proof	301
	C. Signatures	301
	D. Establishing Defenses	302
	E. Establishing Status as Holder in Due Course	303
	F. Third Parties	303

CHAPTER 12. DISCHARGE 306

Section		
1.	Discharge of Underlying Obligation by Issue or Transfer	306
	Example 64	307
	Example 65	308
	Example 66	309
	Example 67	310
	Example 68	311
2.	Discharge of Liability on Instrument—Basic Propositions	311
3.	Discharge as a Defense	312
4.	Bases for Discharge of Liability on Instrument —In General	312
5.	Payment or Satisfaction	313
	Example 69	314
	A. Payment to Owner Who Is Not Holder	315
	B. When Payment to Holder Does Not Effect Discharge	316
	Example 70	317
6.	Tender of Payment	318
	Example 71	318

OUTLINE

Section		Page
7.	Reacquisition	319
	Example 72	319
	Example 73	320
8.	Discharge of Party with No Right of Action or Recourse	320
	Example 74	321
9.	Impairment of Right of Recourse or Collateral	321
	Example 75	322
	Example 76	323
	Example 77	324
10.	Cancellation	325
11.	Renunciation	326
	Example 78	326

CHAPTER 13. CHECKS, CHECK COLLECTION AND RELATIONSHIP BETWEEN BANKS AND THEIR CUSTOMERS—HEREIN OF ARTICLE 4 ... 328

Section		
1.	Overview of the Check Collection Process	329
2.	Rights of Customer in Collection Process	332
	A. Collecting Bank as Agent for Collection	332
	B. Duties of Collecting Bank	332
	C. Provisional Settlement	333
	D. Final Payment	335
	Example 79	339
3.	Contract Between Payor Bank and Checking Account Customer	341
	A. Checking Account Bank Deposit	342
	B. Issuance of Check Is Not an Assignment	343
	C. Certification of Check	343
	Example 80	344
4.	Bank's Liability for Wrongful Dishonor	345
	Example 81	346

Section		Page
5.	Charging Customer's Account	347
	A. Overdraft	347
	B. Drawer's Signature Unauthorized	348
	C. Payment to Non-Holder	348
	Example 82	349
	D. Amount to Be Charged	350
6.	Drawer's General Duty of Care	352
7.	Customer's Duty to Discover and Report Unauthorized Signature or Alteration	352
	Example 83	355
8.	Effect of Death or Incompetence	356
9.	Right to Stop Payment	357
10.	Payment of Stale Checks	359
11.	Bank's Right to Subrogation for Improper Payment	360
	Example 84	362
12.	Summary by Way of Example	363
	A. Forged Drawer's Signature	364
	B. Forged Indorsements	366
	C. Alteration	369

CHAPTER 14. BEYOND COMMERCIAL PAPER: ELECTRONIC FUND TRANSFER SYSTEMS 372

Section		
1.	Why a "Less-Check" Society?	372
2.	Check Handling Systems	374
	A. Magnetic Ink Character Recognition (MICR)	374
	B. Check Truncation	375
	Example 85	376
3.	Electronic Fund Transfer Systems	378
	A. Current Systems	378
	B. Legal Controls	381
4.	A Final Word	383

Index 385

TABLE OF CASES

References are to Pages

American State Bank v. Richendifer, 303
Avery v. Weitz, 171

Bank of Suffolk v. Kite, 90
Bank of Viola v. Nestrick, 60
Batchelder v. Granite Trust Co., 193
Blaine, Gould & Short v. Bourne & Co., 128
Blake v. Coates, 303
Booker v. Everhart, 50, 58, 59

Carpenter v. Payette Valley Co-op, Inc., 146
Central Bank & Trust Co. v. G. F. C., 219
Chemical Bank of Rochester v. Haskell, 250, 253
Cincinnati Ins. Co. v. First Nat'l Bank of Akron, 119, 342
Commonwealth v. _____ (see opposing party)
Corporation Venezolana de Fomento v. Vintero Sales Corp., 85

First Nat'l Bank of Somerset County v. Margulies, 245
First Nat'l Bank of Waukesha v. Motors Acceptance Corp., 244
First State Bank at Gallup v. Clark, 13, 56
Foley v. Hardy, 74
Foutch v. Alexandria Bank & Trust Co., 150
Freeport Bank v. Viemeister, 244

General Inv. Corp. v. Angelini, 274
Gershman v. Adelman, 255
Gorham v. John F. Kennedy College, Inc., 326
Graham v. White-Phillips Co., 254

Kelinson, Commonwealth v., 81
Kirby v. Bergfield, 186

Leonardi v. Chase Nat'l Bank, 131
Liesemer v. Burg, 325
Ligran, Inc. v. Medlawtel, Inc., 173

TABLE OF CASES

McLean v. Bryer, 256
Manufacturers & Traders Trust Co. v. Murphy, 250
Mechanics Nat'l Bank of Worcester v. Killeen, 171
Miller v. Race, 249

Nicholas v. Zimmerman, 301
Northwestern Bank v. Neal, 87

Overton v. Tyler, 56

Peacock v. Rhodes, 104
Price v. Neal, 219, 220, 224, 226, 228
Prudential Ins. Co. v. Marine Bank, 111

Quazzo v. Quazzo, 248

Riley v. First State Bank, 249

Sandler v. Eighth Judicial District Court, 288
Seattle-First Nat'l Bank v. Schriber, 170
Shepherd Mall St. Bank v. Johnson, 121
Snyder v. Town Hill Motors, Inc., 299
Soma v. Handrulis, 130, 131
Stone & Webster Engineering Corp. v. First Nat'l Bank & Trust Co., 366
Sun 'N Sand, Inc. v. United California Bank, 223
Sun Oil Co. v. Redd Auto Sales, Inc., 262

Telpner v. Hogan, 302
Thornton & Co. v. Gwinnett Bank & Trust Co., 115
Trenton Trust Co. v. Western Sur. Co., 286
Trust Co. of Columbus v. Refrigeration Supplies, Inc., 119

Unico v. Owen, 276
Union Constr. Co. v. Beneficial Standard Mortg. Investors, 323
Universal C. I. T. v. Guarantee Bank & Trust Co., 361

Von Frank v. Hershey Nat'l Bank, 78

Weirton S. & L. Co. v. Cortez, 94
West Side Bank v. Marine Nat'l Exchange Bank, 338
Wilmington Trust Co. v. Gesullo, 323
Wilson Supply Co. v. West Artesia, 248

Young v. Grote, 149

TABLE OF REFERENCES TO U.C.C. PROVISIONS AND COMMENTS

Code Sections	Pages
1	6, 7, 35
1-102, Comment 2	11, 74
1-102(1)	373
1-102(3)	11, 74
1-103	10, 11, 294, 346
1-106(1)	346
1-201(1)	144
1-201(2)	46
1-201(14)	85
1-201(15)	10
1-201(19)	46, 249, 273
1-201(20)	44, 84
1-201(24)	10, 64
1-201(25)	216, 225, 252, 254, 258, 272
1-201(25)(c)	254
1-201(26)	252
1-201(27)	253
1-201(28)	144
1-201(30)	52, 144
1-201(31)	302
1-201(37)	41
1-201(43)	145
1-201(44)	239
1-201(44)(d)	165
1-201(46)	51
1-204(2)	255
1-208	66
2	6, 34, 35, 208
2-201(20)	103
2-202	90
2-210	38
2-302	276
2-312	211

TABLE OF REFERENCES TO U.C.C. PROVISIONS

Code Sections	Pages
2-312—2-315	208
2-511	31
3	6, 7, 8, 9, 10, 11, 12, 17, 34, 35, 40, 43, 44, 45, 49, 73, 74, 82, 84, 86, 89, 146, 208, 238, 239, 291, 294, 312, 313, 321, 328, 329, 365, 372, 373
3, Pt. 5	86
3-102(1)(a)	17, 44, 84, 85
3-102(1)(b)	1, 53, 55, 70
3-102(1)(b), Comment 3	55
3-102(1)(c)	53
3-102(1)(d)	151, 175, 184
3-102(1)(e)	9, 82, 238
3-103	8
3-103(1)	10, 34
3-103(2)	10
3-104, Comment 2	75
3-104, Comment 4	21
3-104, Comment 5	69, 75
3-104(1)	2, 9, 11, 44, 49, 50, 51, 74, 77, 238
3-104(1)(a)	52, 83
3-104(1)(b)	9, 51, 52, 55, 56, 62, 64, 73, 78
3-104(1)(c)	65
3-104(1)(d)	69, 80
3-104(2)	8, 10
3-104(2)(a)	1, 17
3-104(2)(b)	19, 328
3-104(2)(c)	1, 22, 53
3-104(2)(d)	14
3-104(3)	82
3-105	51, 55, 56, 74
3-105—3-111	51
3-105—3-112	9
3-105, Comment 7	61
3-105(1)(a)	58
3-105(1)(b)	58
3-105(1)(c)	57, 60, 66
3-105(1)(d)	59
3-105(1)(e)	59
3-105(1)(f)	60
3-105(1)(g)	61

TABLE OF REFERENCES TO U.C.C. PROVISIONS

Code Sections	Pages
3-105(1)(h)	61
3-105(2)(a)	9, 50, 57
3-105(2)(a), Comment 8	59
3-105(2)(b)	60
3-106	62, 74
3-106, Comment 1	63
3-106(1)(a)	62
3-106(1)(b)	62
3-106(1)(c)	62
3-106(1)(d)	63
3-106(1)(e)	63
3-107	64
3-107, Comment 1	64
3-107(1)	64
3-108	65, 171
3-109	65, 74, 171
3-109, Comment 4	66
3-109, Comment 5	67
3-109(1)(a)	66
3-109(1)(b)	66
3-109(1)(c)	66
3-109(1)(d)	67
3-109(2)	68
3-110	46, 69, 101
3-110(1)	70, 75, 80
3-110(1)(a)	71
3-110(1)(b)	71
3-110(1)(d)	71, 119
3-110(1)(e)	71
3-110(1)(f)	71
3-110(1)(g)	71
3-110(3)	73
3-111	69, 72, 101
3-111, Comment 1	72
3-111, Comment 2	72
3-111(a)	101
3-112	51, 73, 77, 78
3-112, Comment 3	75
3-112(1)(b)	51, 59
3-112(1)(d)	78
3-112(1)(e)	77

XLIII

TABLE OF REFERENCES TO U.C.C. PROVISIONS

Code Sections	Pages
3-113, Comment	294
3-114, Comment 2	81
3-114(1)	80, 81
3-114(2)	81
3-115	81
3-115(1)	80, 152
3-116(a)	119
3-116(b)	119
3-118	51, 88, 89, 171
3-118, Comment 1	91, 93
3-118, Comment 7	67
3-118(a)	54
3-118(b)	92
3-118(b), Comment 6	73
3-118(c)	92
3-118(d)	63, 173
3-118(f)	67, 68
3-119	57
3-119, Comment 5	50
3-119(1)	56, 87
3-119(2)	50, 57, 87
3-121	195
3-122	86, 169, 171, 173
3-122(1)(a)	171, 183, 190
3-122(1)(b)	170, 191
3-122(2)	172
3-122(3)	171
3-122(4)	173
3-201	86
3-201(1)	99, 233, 237, 266, 268, 313
3-201(3)	117, 159
3-201(3), Comment 7	117
3-201(20)	99
3-202, Comment 4	120
3-202(1)	46, 73, 85, 99, 100, 101, 123, 367
3-202(1)(c)	264
3-202(2)	102, 104, 121
3-202(3)	120
3-202(4)	136
3-203	121

TABLE OF REFERENCES TO U.C.C. PROVISIONS

Code Sections	Pages
3–204(1)	46, 102, 107
3–204(2)	101, 102, 104
3–204(3)	107
3–205	123
3–205(c)	269, 332
3–206(1)	124, 125, 332
3–206(2)	125, 130, 131
3–206(3)	125, 127, 131, 269, 270, 332
3–206(4)	133, 264
3–207, Comment 1	122
3–207, Comment 2	122
3–207(1)	121
3–208	319
3–301	9, 44, 86, 99
3–302	127, 270, 272, 273
3–302, Comment 2	265
3–302(1)	46, 238
3–302(1)(a)	240, 274
3–302(1)(c)	45, 252, 254
3–302(3)	267
3–303	46, 239, 240, 241
3–303, Comment 6	245
3–303(a)	165, 243, 248
3–303(b)	247, 248
3–303(c)	243
3–304	46, 252
3–304, Comment 2	263
3–304, Comment 6	257
3–304, Comment 10	262
3–304, Comment 12	253
3–304(1)(a)	261
3–304(1)(b)	260
3–304(2)	263
3–304(3)(a)	257
3–304(3)(b)	256
3–304(3)(c)	255
3–304(4)	252
3–304(4)(a)	262
3–304(4)(b)	261
3–304(4)(d)	262
3–304(4)(e)	263

TABLE OF REFERENCES TO U.C.C. PROVISIONS

Code Sections	Pages
3-304(4)(f)	257
3-304(6)	253
3-304(7)	253
3-305	45, 46, 49
3-305(1)	122, 237, 298, 299
3-305(2)	45, 46, 122, 237, 265, 270, 274, 285, 291, 304
3-305(2)(d)	290
3-306	9, 46, 86, 304
3-306(a)	299
3-306(b)	45, 293
3-306(c)	240, 295
3-306(d)	297
3-307	9, 28, 45, 86, 100, 277, 278, 304
3-307(1)	301
3-307(1)(a)	52, 302
3-307(1)(b)	142, 302
3-307(2)	86, 302
3-307(3)	303
3-401	52
3-401, Comment 2	52
3-401(1)	83, 101, 102, 138, 142
3-402	120, 140, 172
3-402, Comment	140, 141
3-403(1)	139, 142
3-403(2)(a)	138, 142
3-403(2)(b)	140, 143
3-404, Comment 3	146
3-404(1)	139, 146
3-405	116, 365
3-405, Comment 1	116
3-405, Comment 2	115
3-405, Comment 4	111
3-405(1)(a)	114
3-405(1)(b)	109, 110
3-405(1)(c)	110
3-406	139, 149, 150, 151, 227, 292, 352, 366, 368
3-407	215, 375
3-407(1)(b)	80, 147
3-407(2)(a)	80, 147

TABLE OF REFERENCES TO U.C.C. PROVISIONS

Code Sections	Pages
3-407(2)(b)	147
3-407(3)	147, 230
3-408	86, 240, 270, 293, 294, 301
3-408, Comment 2	294
3-409, Comment 1	155
3-409(1)	153, 154, 343
3-410(1)	20, 27, 153, 156
3-410(1), Comment 5	156
3-410(3)	156
3-411(1)	20, 343, 344
3-411(1), Comment 1	344
3-411(2)	343
3-412(1)	158
3-412(2)	195
3-412(3)	158
3-413(1)	15, 20, 44, 53, 70, 152, 154, 156, 170
3-413(2)	15, 54, 70, 153, 171, 175, 179, 184
3-414, Comment 1	123, 161
3-414, Comment 4	161
3-414(1)	46, 101, 122, 159, 160, 161, 171, 172, 175, 184
3-414(2)	122, 160
3-415	166
3-415, Comment 1	166
3-415, Comment 3	294
3-415(1)	164, 173
3-415(2)	164, 165, 240
3-415(5)	164, 166
3-416(1)	168, 172
3-416(2)	168
3-416(3)	168
3-416(4)	172
3-416(5)	168, 176
3-417	123, 235
3-417, Comment 3	224, 368
3-417, Comment 9	218
3-417(1)	208, 209, 221, 222, 232, 330, 339, 348
3-417(1)(a)	222

TABLE OF REFERENCES TO U.C.C. PROVISIONS

Code Sections	Pages
3-417(1)(b)	225, 237
3-417(1)(b)(i)	226
3-417(1)(b)(iii)	227
3-417(1)(c)	229, 237
3-417(1)(c)(i)	229
3-417(1)(c)(ii)	229
3-417(1)(c)(iii)	230
3-417(1)(c)(iv)	230
3-417(2)	46, 208, 209, 240, 305
3-417(2)(a)	210, 222
3-417(2)(b)	212
3-417(2)(c)	215
3-417(2)(d)	215
3-417(2)(e)	217
3-417(3)	162, 216
3-417(4)	235
3-418	220, 221, 330, 339, 348, 365
3-418, Comment 1	220, 365
3-419	11
3-419(1)	119
3-419(1)(a)	155
3-419(1)(b)	155, 163
3-419(1)(c)	163, 349, 368
3-419(2)	368
3-419(3)	365
3-501, Comment 2	175
3-501(1)(a)	175, 189, 195
3-501(1)(b)	175, 176, 184, 195
3-501(1)(c)	175, 184, 195
3-501(2)(a)	175
3-501(2)(b)	175
3-501(3)	201
3-501(4)	176
3-502(1)(a)	175, 176, 179, 184
3-502(1)(b)	175, 176, 181, 184
3-502(1)(c)	179
3-502(2)	176, 179, 182, 201
3-503, Comment 1	190
3-503(1)	11
3-503(1)(a)	189
3-503(1)(b)	189

TABLE OF REFERENCES TO U.C.C. PROVISIONS

Code Sections	Pages
3-503(1)(c)	189, 193
3-503(1)(d)	192
3-503(1)(e)	189, 191
3-503(2)	191, 192, 329
3-503(3)	190
3-503(4)	188
3-504, Comment 1	186
3-504(1)	184, 185
3-504(2)(a)	186
3-504(2)(b)	186
3-504(2)(c)	205, 329
3-504(3)(a)	187
3-504(3)(a), Comment 3	55
3-504(4)	187
3-505(1)	187, 329
3-505(2)	187
3-506(1)	11, 196
3-506(2)	196
3-507(1)(a)	195, 257, 258
3-507(1)(b)	197, 258
3-507(2)	175, 197
3-508(1)	198, 199
3-508(2)	200
3-508(3)	197, 198
3-508(4)	197
3-508(5)	198
3-508(6)	198
3-508(7)	198
3-508(8)	199
3-509, Comment 4	200, 201
3-509, Comment 6	202
3-509, Comment 7	201
3-509(1)	200
3-509(2)	200
3-509(4)	201
3-510, Comment 2	196
3-510(a)	202
3-510(b)	203
3-510(c)	203
3-511	204

TABLE OF REFERENCES TO U.C.C. PROVISIONS

Code Sections	Pages
3-601(2)	312
3-601(3)(a)	320
3-601(3)(b)	321
3-602	260, 312, 320
3-603	370
3-603(1)	224, 316
3-603(1)(b)	125, 135
3-604(1)	194, 318
3-604(2)	318
3-604(3)	194, 319
3-605(1)	326
3-605(1)(a)	325
3-605(1)(b)	326
3-605(2)	325
3-606(1)(a)	322
3-606(1)(b)	324
3-606(2)	321
3-606(2)(a)	323
3-606(2)(c)	323
3-802(1)	44
3-802(1)(a)	308
3-802(1)(b)	45, 86, 175, 307, 310
3-803	304
3-804	300, 301, 315
3-805	9, 10, 76, 238
4	6, 7, 8, 10, 27, 34, 208, 328, 329, 330, 372, 373, 374, 375
4-102(1)	10, 328
4-103	374
4-103(1)	330, 332, 342, 359
4-103(2)	330
4-103(5)	330
4-104(1)(e)	333
4-104(1)(g)	328, 330
4-104(1)(h)	328, 365
4-105	200, 330, 334
4-105	126
4-105(a)	269
4-105(b)	269, 328
4-105(c)	330
4-105(d)	269, 328, 330
4-108	374

TABLE OF REFERENCES TO U.C.C. PROVISIONS

Code Sections	Pages
4-109	331, 336, 338, 374
4-109(e)	337
4-201(1)	270, 332
4-201(2)	132, 332
4-202(1)	332
4-202(1)(b)	333
4-202(1)(c)	333
4-202(2)	333
4-204	374
4-204(1)	333
4-204(2)	333
4-205	332
4-205(1)	118
4-205(2)	130
4-207	208, 235, 236, 375
4-207(1)	235, 330, 331, 339, 365
4-207(1)(a)	365, 368
4-207(1)(b)	348, 365
4-207(1)(c)	370
4-207(2)	235
4-207(3)	209
4-208	333, 334
4-208(1)(a)	245, 270
4-208(1)(b)	247
4-208(2)	246
4-209	246, 333
4-210(1)	193
4-211(1)	334
4-212(1)	334
4-213	338
4-213(1)	331, 335, 364
4-213(1)(a)	330
4-213(1)(c)	331, 337
4-213(1)(d)	334, 335
4-213(3)	333, 334
4-301	258
4-301, Comment 1	334
4-301(1)	334, 335
4-301(1), Comment 1	331
4-302(a)	329, 331, 334
4-303	337, 338

TABLE OF REFERENCES TO U.C.C. PROVISIONS

Code Sections	Pages
4–303(1)	338, 359
4–303(1)(d)	338
4–401	328, 347, 375
4–401, Comment 1	347
4–401(1)	109, 154, 342, 347, 363, 364
4–401(2)	351, 370
4–401(2)(a)	370
4–401(2)(b)	369
4–402	345, 346
4–403, Comment 3	358
4–403(1)	357, 358
4–403(2)	358
4–403(3)	367
4–404	360
4–405, Comment 2	357
4–405, Comment 3	357
4–405(1)	356
4–405(2)	356, 358
4–406	375, 376
4–406, Comment 3	354
4–406(1)	353, 366
4–406(2)(a)	354
4–406(2)(b)	354
4–406(3)	355, 366
4–406(4)	355
4–407	349, 360, 364, 368
4–407(a)	361
4–407(b)	362, 367, 369
4–407(c)	362, 366
4–501—4–504	29
5	6, 7, 34, 280
6	7
7	7
7, Pt. 5	34
8	7, 25, 373
8–102	10
8–321	25
9	7, 10, 12, 25, 41, 42, 43
Art. 9, Part 5	35
9–102(1)	41
9–102(2)	41

LII

TABLE OF REFERENCES TO U.C.C. PROVISIONS

Code Sections	Pages
9–104(f)	41
9–105(1)(*l*)	41
9–106	12, 41
9–201	42
9–203	41
9–206	272, 273, 274
9–206(1)	75, 268, 269, 276
9–301	42
9–302(1)	42
9–302(1)(e)	42
9–303	42
9–303(1)	41
9–312(5)	43
9–318	39, 42, 43
9–318(1)	42
9–318(4)	12, 124
9–401	42
9–402	42
10	7
11	7
16, Pt. 433	279

UNIFORM CONSUMER CREDIT CODE

3.404(1)	277
3.404(2)	278
3.404(4)	278
3.405(1)	279
3.405(1)(e)	278

*

COMMERCIAL PAPER
IN A NUTSHELL

*

CHAPTER 1

INTRODUCTION

§ 1. Importance of Commercial Paper

Drafts, checks and promissory notes—the principal kinds of commercial paper—perform important business functions. Suppose that M, a merchant, owes B $1,000 from a previous transaction and the debt is due on June 1. On May 15, B agrees to buy goods from C for $1,000 and C demands payment by June 1. If M is known and trusted, C might agree to take a *draft* drawn by B on M as conditional payment for the goods. By a *draft* (or bill or exchange), B (Drawer) signs a writing ordering M (Drawee) to pay C (Payee) $1,000 on June 1 when the writing is presented. If M pays C upon presentment, the *draft* has facilitated the discharge of M's debt to B and B's debt to C and enabled B to buy goods from C. In this case, M was a merchant and B drew what is called a "time" *draft*. If M were a bank and the *draft* was payable on demand, the writing ordering M to pay C would be called a *check*. See Sections 3–104(2)(a) and (c) and 3–102(1)(b) of the Uniform Commercial Code. In either case, the draft enables the drawer to transfer funds in payment of obligations.

Suppose, instead, that C refused to take B's draft on M as payment. Rather, he agreed to give B until August 1, to pay. B's promise to pay C $1,000 on August 1 is enforceable as a contract. C's right to payment is an intangible form of personal property which can be transferred or assigned to third parties. But, as we shall see, the promise might be more valuable to C if expressed in a writing and in a form that qualifies it as a *negotiable promissory note*. Thus, if B signed a writing promising unconditionally to pay C "or his order" $1,000 on August 1 and delivered the writing to C, C would be the holder of a *negotiable promissory note*, i. e., a written promise to pay that satisfies the requirements of form in Section 3–104(1) of the Code. Of what advantage is this to C? Suppose C owed D $950 and this was due on June 1. D might be willing to take C's note for $1,000 in full payment of the debt. If all goes well, on August 1 D will present the note to and demand payment from B (Maker) and B will pay. The note, therefore, has performed important credit functions. As reliable evidence of B's debt to C (Payee), it facilitated the transfer from C to D (Holder) in satisfaction of C's debt and enabled D to profit from the difference between the face amount of the note and the amount of C's debt. B needed the credit and since C had delivered the goods, B will be indifferent to whether he must pay C or D on August 1. Finally, because the note is negotiable and was negotiated to C, who took it for value, in good faith and without notice of any defects or defenses, D will be a holder in due course. This important status protects D from most claims and defenses that B (Maker) will have against C (Payee) when the time for payment arrives. We will have much more to say about the holder in due course.

For a fuller discussion of the uses of commercial paper, see Chapter 2, Section 2.

There are risks in these transactions. In the case of the *draft*, M might not follow B's order to pay C. What then? In the case of the *note*, B might refuse to pay D, the holder, because the goods delivered by C (Payee) were defective. What then? Is D a holder in due course and how can we tell? When disputes arise under writings devised by private parties to facilitate business activity and cannot be settled by agreement, there is a need for law. The body of Anglo-American law that has evolved over the last 800 years in response to these and other questions is called the law of commercial paper.

§ 2. Early Law of Commercial Paper

Professor Britton has stated that by the "close of the 1700's * * * the basic principles of the law of negotiable instruments had been laid down in the decisions of the English courts." Britton, Bills and Notes 9 (2d ed. 1961). From then until codification of the law of commercial paper in the late 19th Century, the courts amplified and applied these principles in response to the demands of a growing and complex economy.

Before 1702, however, the common law development had favored the draft or "bill or exchange" over the promissory note. The "bill" was used by domestic and foreign merchants in England as early as 1300 to exchange foreign currency, transfer domestic funds and finance the sale of goods. Various principles of liability and remedy and the concept that a bill was freely transferable were developed for merchants in the mercantile courts prior to 1600 and in the common law courts thereafter. By the end of the 17th Century, these principles had been applied to all parties on a bill of exchange, whether merchants or not. By the end of the 18th Century, the common law, aided by Lord Mansfield, had begun to explore the rights of good faith purchasers and the relationship between the bill and the underlying obligation. Clearly, the customary use and commercial importance of the bill of exchange eased its acceptance and development in the common law after the triumph of assumpsit in 1602.

It was another story for promissory notes. Notes, then called "bills obligatory," were used before 1600. Their free transferability, however, was not accepted by the common law. English courts, along with continental and Roman law, did not recognize the concept that the owner of an intangible promise could transfer it to a third person without the delivery of some tangible token. Delivery of a simple writing was not enough. Fictions were developed to circumvent this limitation, but it was not until the Statute of Anne in 1702 that notes were made "assignable or indorseable over" in the same manner as bills of exchange. Once the concept of free transferability was recognized for the note, the common law courts were then able to recognize and develop its key function as a facilitator of credit transactions.

§ 3. Uniform Negotiable Instruments Law

In 1896, the National Conference of Commissioners on Uniform State Laws promulgated the Uniform Negotiable Instruments Act, which is often referred to as the Negotiable Instruments Law or the N.I.L. Like the British Bills of Exchange Act (1882), after which it was in most respects modeled, the N.I.L. was largely declaratory of the common law. By 1924 it had been adopted with only minor variations in all states.

§ 4. Uniform Commercial Code

Eventually, the N.I.L. was found to contain serious defects which resulted primarily from its failure to provide for changing commercial practices. Also, differences in interpretation among the courts tended to destroy the uniformity which it was intended to provide.

Similar weaknesses appeared in other areas of commercial law. Some legal authorities felt that these difficulties should be overcome by piecemeal amendments of existing statutes; others thought that a comprehensive statute to deal with the various areas of the law governing commerce was necessary. The efforts of the latter group during several decades resulted in the Uniform Commercial Code—usually referred to as the "Code."

The Code was primarily the joint work of the American Law Institute and the National Conference of Commissioners on Uniform State Laws. In preparing the Code, these organizations used the knowledge, experience, and talents of literally hundreds of judges, lawyers, law professors, bankers, merchants, and many other persons from various parts of the country. However, the 1952 Official Text, the first, was adopted in only one state, Pennsylvania. It was not adopted in other states largely because the legislature of New York, a key state in commercial matters, referred the Code to its Law Revision Commission which, after making an intensive study over a period of several years, reported to the New York Legislature in 1956 that "the Code is not yet ready for enactment." (Legislative Document 1956 No. 65(A) 5.)

The New York Law Revision Commission Report contained a large number of detailed criticisms and suggestions. After studying these, the Code editorial board recommended extensive amendments of the 1952 Official Text, which were reflected in the 1957 Official Text. Many of these amendments related directly to the provisions treating commercial paper. Later Official Texts (1958, 1962, 1972, 1978), however, made few significant changes in the Code provisions relating to commercial paper.

As of January 1, 1982, the Code has been adopted so as to be effective in all states, except Louisiana (which has adopted Articles 1, 3, 4, 5), as well as in the District of Columbia and the Virgin Islands. Thus far, although Congress has enacted the Code for the District of Columbia, it has refused to adopt it as general federal statutory law.

The Code deals with not only commercial paper, which is covered in Article 3, but also with sales of goods (Article 2), bank deposits and collections (Article

4), letters of credit (Article 5), bulk transfers (Article 6), documents of title (Article 7), investment securities (Article 8), and secured transactions (Article 9). Article 1 contains a number of general provisions including a large number of general definitions and general rules of construction and principles of interpretation. Article 10 provides the effective date of the Code and also repeals the Negotiable Instruments Law and a number of other state laws. Article 11, which was added by the 1972 Official Text, is a transition provision that indicates when the original version of Article 9 continues to govern secured transactions in states in which that Article has been substantially revised after the particular secured transaction was consummated by the parties.

The following discussion is based primarily on Article 3, entitled "Commercial Paper," which has earned the approval of almost all legal scholars, the courts, and lawyers, as well as bankers and other businessmen. There are several reasons why Article 3 serves well as a basis for a discussion of the law of commercial paper. One reason is that it is basically a restatement, clarification, and modernization of such law, and embodies and retains virtually all of the sound principles of its predecessor, the N.I.L., and of the cases decided before and after the adoption of the N.I.L. Another reason is that in most situations in which the courts were divided under the prior law, the Code ended the division by adopting one view or the other, or a third view. Also, the Code, to a far greater degree than almost any other comprehensive statute, was carefully thought out and anticipated possible problems of interpretation as well as deviations from normal business conduct. Finally, the Code dealt with these problems and deviations directly and expressly rather than leaving them to be decided by the courts in the painful process of litigation.

Nevertheless, as the Third Edition of this Nutshell goes to press, an extensive study of the suitability of Articles 3 and 4 in the age of the computer and electronic fund transfers is nearing completion. Without question, proposals for the revision of these Articles will be made in the years ahead. See, e. g., Leary and Tarlow, *Reflections on Articles 3 and 4 for A Review Committee*, 48 Temp.L.Q. 919 (1975).

§ 5. The Scope of UCC Article 3

Article 3 is but one segment in a commercial code which, in turn, is part of a complex and changing pattern of state and federal law. There are many overlaps and conflicts in these patterns. Article 3, therefore, is no more the exclusive statement of commercial paper law than the Code is the exclusive statement of general commercial law. We must, however, have a firm notion about what Article 3 does and does not cover. To obtain this, a second encounter with the concept of negotiability is required.

A. INSTRUMENTS ARE WITHIN SCOPE

Article 3 deals, in general, with commercial paper, which includes writings called drafts and notes. 3–104(2). Section 3–103 lists some situations where Article 3 does not apply but there is no section which states to what Article 3 does apply. The conclusion that it applies to transactions involving instruments must be derived from the overall structure.

An "instrument" is defined in Section 3–102(1)(e) as a "negotiable" instrument and the conditions for negotiability are set forth in Section 3–104(1). Read that section, please. A note, for example, is negotiable if and only if it is "signed by the maker," contains an "unconditional promise * * * to pay a sum certain in money," is payable at a "definite time" and is payable to "order or bearer." These conditions are elaborated in Sections 3–105 through 3–112. Clearly, Article 3 is always applicable to determine whether a writing is an instrument. If the writing is an instrument, the full scope of Article 3 is applicable. Article 3 also applies to a writing which is otherwise negotiable but is not payable to "order or to bearer." 3–805. But Article 3 is not applicable in other cases where the writing fails to qualify as a negotiable instrument. To illustrate, suppose that Maker promises to pay "Payee or order" $5,000 on November 1 "if the Chicago Cubs win the World Series." Payee negotiates the note to D for value. The Cubs do win, D presents the note to Maker for payment and is told that the note was issued to Payee as part of an illegal wager. Article 3 is applicable to determine whether the note was negotiable. Clearly, it was not because the promise was conditional. 3–104(1)(b) and 3–105(2)(a). D's rights as possessor of the note, however, are not governed by Article 3. D must be the "holder of an instrument" before he has rights under Article 3, 3–301, 3–306 and 3–307. If the writing is not an instrument, the general law of contracts would determine the rights and duties of the parties. See 1–103. On the other hand, if the note contained an unconditional promise to pay Payee but omitted the words "or order," the writing, although not negotiable, would be treated as an instrument under Section 3–805, except that Payee could not be a holder in due course. His other rights and duties, however, would be governed by Article 3.

B. EXCLUSIONS FROM SCOPE

Assume that our note is an instrument. Are there any other limitations on the applicability of Article 3? The fact that the writing is a note insures that some potential limitations are not applicable. The note is neither money, 1–201(24), nor a document of title, 1–201(15), nor an investment security, 8–102. If it were, Article 3 would not apply even though the writing was negotiable. See 3–103(1). Similarly, Article 4 is not applicable because a note is not a check, see 3–104(2), and banks and bank collections will not be involved. See 4–102(1). Finally, Article 9 will not apply unless the note becomes collateral for a loan and a security interest in it is created and perfected. See 3–103(2). For now, note the possibility that an instrument governed by Article 3 *may*, in certain circumstances, be covered by sections in Articles 4 and 9. To the extent that Articles 4 and 9 apply to any particular transaction, they prevail over Article 3. 3–103(2).

Three other circumstances may limit the scope of Article 3. First, many states have enacted legislation eliminating the concept of negotiability in consumer credit transactions. The holder of a consumer note is subject to Maker's claims and defenses. Similarly, the Federal Trade Commission has promulgated a rule designed to accomplish the same objective. See the discussion in Chapter 10, Section 4. Legislation of this sort preempts conflicting provisions of Article 3. Second, the particular provisions of Article 3 may fail sufficiently to displace other "principles of law and equity, including the law merchant * * *" otherwise applicable in the state. Unless displaced by the "particular provisions" of Article 3, these principles shall supplement the law of commercial paper. 1–103. A good example is the law of conversion, covered only in part by Section 3–419. Finally, the parties have a limited power to alter or vary the effect of Article 3 by agreement. 1–102(3). This power is limited by the nature of Article 3, which prescribes the formalities necessary for negotiability and protects the transferability of commercial paper. Thus, an agreement can not make negotiable a writing that does not comply with Section 3–104(1), or change the meaning or rights of a holder in due course. Comment 2, 1–102. Agreement might, however, define the time for presentment of an instrument, 3–503(1), or the time allowed for acceptance or payment, 3–506(1).

Enough has been said to suggest that scope questions can be tricky and require constant attention. Consider the following examples.

Example 1

Bank issued a certificate of deposit payable to Customer (but not "to order") and containing a restriction that it could not be transferred without Bank's written consent. Customer, without obtaining that consent, assigned the CD to D as security for a loan. Twelve months later, Customer defaulted on the note to D and, when the CD became due, D presented it to Bank for payment. In the meantime, defenses and claims arising out of a separate transaction between Customer and Bank had arisen. Bank asserted that because of the restriction D was not entitled to enforce the CD and that, in any event, D was subject to the claims and defenses.

The CD was not negotiable, 3–104(1), and was not, therefore, an instrument. Further, it was not "otherwise negotiable" within the meaning of Section 3–805. Thus, D could not be a holder in due course and Bank could not use Section 3–306 as the basis for asserting claims and defenses. Article 3 was not applicable after the question of negotiability was decided. Thus, the source of Bank's setoff rights, if any, must be the common law rather than the Code. See Restatement, Second, Contracts §§ 322 and 336.

D could argue that Article 9 rather than the common law governed the transaction. D is the owner of a CD in which a security interest was created by assignment and Article 9 applies in general to this transaction. 9–102(1). If D is correct, Section 9–318 governs the enforceability of the restriction on transfer, 9–318(4), and whether Bank can assert claims and defenses arising from transactions with Customer against D, 9–318(1). On these facts, D is correct. The CD is a "general intangible" rather than an instrument under Article 9's classification system, see 9–105(1)(i) and 9–106, and Bank is an "account debtor" within the meaning of 9–318(1). See 9–105(1)(a). Thus, Section 9–318 rather than the common law is the source of Bank's right to setoff against D.

Example 2

Maker issued to Payee an otherwise negotiable note which contained on its face the following language: "This negotiable note shall not be transferred without Maker's consent." When Payee attempted to transfer the note to D, D telephoned Maker and inquired about the restriction. Maker replied: "There is nothing wrong here. Go ahead and take the note." D took the note for value, in good faith and without notice of any defenses between Maker and Payee. Later, defenses inherent in the transaction at the time of transfer arose and Maker asserted them against D. The restriction impaired negotiability and the statement that the note was "negotiable" was not effective. Maker, however, consented to the transfer and may be estopped to raise any defenses of which D was unaware. A court may arrive at a result similar to negotiability in a particular case based upon principles of estoppel. See Comment 2, 3–104, 1–103 and First State Bank at Gallup v. Clark, 91 N.M. 117, 570 P.2d 1144 (1977).

CHAPTER 2

FORMS AND USES OF COMMERCIAL PAPER AND TERMINOLOGY

It is helpful in acquiring a sound understanding of the legal principles governing commercial paper to learn to recognize quickly the various forms of such paper and the different capacities in which the names of the parties may appear on it, and to become acquainted with the different uses of each kind of instrument. In Chapter 2, therefore, we elaborate on some ideas introduced in Chapter 1, Section 2.

§ 1. Forms of Commercial Paper and Parties

A. PROMISSORY NOTE

The most elementary form of commercial paper is the *promissory note*, which usually is referred to simply as a *note*. A note is a written promise other than a certificate of deposit (see page 32) to pay money signed by the person making the promise, who is called the *maker*. 3–104(2)(d). In the examples that appear throughout this book the maker is usually iden-

COMMERCIAL PAPER Ch. 2

tified by a capital "M." In banking circles and in some courts the term "maker" commonly refers to the drawer of a check. But in the Code and this book "maker" refers exclusively to one who signs a note or a certificate of deposit. Like other forms of commercial paper, a note may be made payable to a named person, to his order, or to bearer. If a note or any other form of instrument when first issued is made payable to a named person or to his order, he is called the *payee*. In the examples mentioned above the payee usually is identified by a capital "P."

The promissory note is the only kind of commercial paper that undertakes to express the obligation of the party who issues it. See 3–413(1). As we shall see later, the obligation of the drawer of a check or draft is not so expressly revealed and neither is the obligation of the maker of a certificate of deposit. See 3–413(2).

A note may take any of a number of different forms, depending on its purpose, the understanding of the parties and other factors. A simple note might take the following form.

$5,000.00   New York, N.Y.   July 1, 1981.

One year after date I promise to pay to the order of Paul Payee five thousand dollars.

Due July 1, 1982   (Signed) Michael Maker

Ch. 2 *FORMS AND USES; TERMINOLOGY*

The above note is payable at a definite time. A note might instead be payable on demand. (See pages 65–66.) It is common to refer to a note that is payable at a definite time as a *time* note or time instrument and to refer to one that is payable on demand as a *demand* note or demand instrument. If an instrument is payable on demand, the holder normally is entitled to demand payment at any time after it is issued. If a demand instrument is post-dated, however, the holder must wait until the date shown on the instrument arrives.

Usually, a note is prepared on a printed form which originally contained blank spaces for the date of issue, the place, the amount, the name of the payee, and the signature of the maker. If the payee is a bank or other lending institution, the name of the payee is likely to appear on the original printed form. Typically, the form also contains the words "value received" and "with interest at the rate of" and leaves blanks for the interest rate, the due date, the place of payment, the number of the note, a memorandum of the transaction, the address of the maker and for other details. A note might also contain provisions relating to security, payment in instalments, acceleration, renewals, other notes in the same series, and many other matters. Very little of what might be included is necessary for a valid and enforceable note. An instrument that provides, "I will pay John Jones $1,000," may be enforced as a non-negotiable note if it is signed by a maker.

(For a wide range of examples of forms that commercial paper might take under Article 3, along with useful commentary, see Uniform Laws Annotated, Uniform Commercial Code, Master Edition, Vol. 4 Forms, Henson & Davenport, West Publishing Company 1968, with Annual Supplement.)

B. DRAFT

Draft and *bill of exchange* are synonyms. 3–104(2)(a). However, since the Code uses *draft*, and since *bill of exchange* is rarely used except in international trade, we will use the word *draft*.

A draft is an *order* to pay money. The person issuing the order is the *drawer* and the person to whom the order is addressed is the *drawee*. In the examples, the drawer usually is identified by a capital "R," and the drawee is usually identified by a capital "E." In virtually all drafts, the order to pay is expressed by the word "pay." See 3–102(1)(a). A draft is basically different from a note because it does not contain a promise. The drawer of a draft does not expressly promise to pay it. Indeed his normal expectation is that he will not have to pay it. He orders the drawee to pay. However, if the drawee does not pay, the drawer, under circumstances that will be discussed later, may be required to pay. Like a note, a draft may be payable to bearer or to some specific person or to

Ch. 2   *FORMS AND USES; TERMINOLOGY*

his order. Sometimes the drawer names himself as payee, but normally he names a third party. Like a note, a draft may be payable on demand or at a definite future time. (See Chapter 4, Section 1G.)

A simple draft might take the following form.

Miami, Florida   May 1, 1981

One year from date pay to the order of Payee & Company

$20,000.00   Twenty thousand.......... Dollars.

(Signed) David Drawer
DAVID DRAWER

To: DANIEL DRAWEE
   10 Broad Street
   Rome, Georgia

Like a note, a draft usually is prepared on a printed form. Although a draft might contain many terms and furnish much information, most drafts are simple in form.

C. CHECK

By far the most widely used form of commercial paper is the check. Each year the Federal Reserve

Banks alone process about ten billion checks that total about five trillion dollars.

A check is a *draft* which is drawn by the drawer on a *bank*, as drawee, and is payable *on demand*. 3–104(2)(b). All other forms of commercial paper may be payable either on demand (for example, a demand note) or at some definite future time (for example, a 60-day note). Although a check is one kind of draft, it is common to refer to a draft that is drawn on a bank and is payable on demand as a "check"; and it is common to refer to any other draft simply as a "draft", although occasionally in the interest of precision, it is called an "ordinary draft" or a "non-check draft." When the word "draft" is used without qualification in the statement of legal principles, however, the term includes checks as well as other kinds of drafts unless the context indicates otherwise.

Below is a simple form of check.

September 15, 1981

Pay to the order of Paul Payee $1,000.00

One Thousand........................ Dollars

ALLINE BANK   DRAWER CORP.

Alline, Ohio   By (Signed) Daniel Drawer,
                                    Treasurer

Ch. 2 *FORMS AND USES; TERMINOLOGY*

To facilitate passage through banking channels, however, most checks contain also, in the upper right hand corner a hyphenated number over another number. Most checks contain, in addition in the lower lefthand corner, a combination of numbers and symbols which are required by the Magnetic Ink Character Recognition check collection system pursuant to which checks are passed through a computer which sorts the checks and charges each check to its proper account. In addition, for the convenience of the depositor, most checks that are furnished by banks contain the customer's name and address printed on the face, a space for the number assigned to the check by the customer, and a space for recording the reason for issuing the check. See Speidel, Summers & White, Commercial and Consumer Law 1389–1397 (3d ed. 1981). In contrast with notes, the vast majority of checks show little variation with respect to the provisions that affect the rights and duties of the parties.

The drawee of a check, like a non-bank draft, is not liable on the instrument unless it accepts. One form of acceptance is called certification. Either Drawer or Payee obtains Drawee Bank's signature on the face of the instrument. See 3–411(1) and 3–410(1). Drawee Bank then becomes primarily liable on the instrument. 3–413(1). Another variation is the so-called "cashier's check." Suppose the buyer of goods has an account with Bank and Seller insists upon payment by "cashier's check." If there are sufficient funds in the account, Bank will draw a check against itself to Seller or order. Put another way, an officer of Bank will sign the check as Drawer ordering itself to pay Seller or order. Buyer does not sign the instrument in any capacity and the legal relationship on the check is between Seller (Payee) and Bank (both Drawer and Drawee). Buyer's account, however, has been debited by Bank.

One further variation is the so-called "traveler's check." Although intended to be within Article Three, Comment 4, 3–104, this instrument is difficult to classify. Is it money or a certificate of deposit or a cashier's check or what? Professor Hawkland has argued that it should be treated as a cashier's check. The issuer is both Drawer and Drawee and the purchaser is Payee. The purchaser's countersignature is, in effect, an indorsement and the party who "cashes" the check, be it bank or retailer, is a holder by negotiation. See Hawkland, *American Traveler's Checks*, 84 Banking L.J. 377 (1967). The classification issue becomes important when the checks are lost or stolen, but its practical import is diminished by the policy of most issuers to reimburse the purchaser.

D. CERTIFICATE OF DEPOSIT

A certificate of deposit, often referred to as a "CD," is the acknowledgement by a *bank* of a receipt of mon-

Ch. 2 *FORMS AND USES; TERMINOLOGY*

ey and an express or implied promise to repay it. 3–104(2)(c). The following is typical.

NEGOTIABLE TIME CERTIFICATE
OF DEPOSIT

FIRST CITY BANK OF ARIZONA

No. 5693   Tuscon, Arizona   June 1, 1981

THIS CERTIFIES THAT THERE HAS BEEN DEPOSITED with the undersigned the sum of $200,000.00

Two hundred thousand................. Dollars

Payable to the order of PAYEE DEPOSITOR CORPORATION on December 1, 1983 with interest only to maturity at the rate of FIFTEEN per cent (15%) per annum upon surrender of this certificate properly indorsed.

FIRST CITY BANK OF ARIZONA

By (Signature) Dirk Duer, Vice-President

Authorized Signature

Thus, it resembles an ordinary note in that the party who issues it promises to pay; and it resembles a check in that the person who is expected to pay is always a bank. The bank that issues a certificate of deposit may properly be referred to as the "maker." A certificate of deposit may take any one of a number of forms.

§ 2. Uses of Commercial Paper

A. PROMISSORY NOTES

A promissory note may be used for many different purposes, but generally these purposes fall into one of two important categories: (1) a means of borrowing money, and (2) a means of buying on credit. A third use is as a method of evidencing a pre-existing debt.

1. MEANS OF BORROWING MONEY

The first purpose mentioned is probably the most obvious. For example, assume that a businessman borrows $50,000 from a bank to help finance the purchase of raw materials. As evidence of the loan, he is required to sign and deliver to the bank a negotiable note in which he promises to pay to the order of the bank six months after date the sum of $50,000.

The bank computes the amount of interest on $50,000 for six months at the agreed rate, deducts (or discounts) this amount from the $50,000 and places the balance as a credit in the maker's checking account. The bank may hold this note until it is paid, or rediscount it at another bank where a lower rate of interest is charged, thus obtaining cash or credit and netting a profit for the difference between the two rates of discount.

In order to induce a second bank to rediscount the note, the payee bank will have to *indorse (or endorse)* it, so as to assume secondary liability. (The indorser's

Ch. 2 *FORMS AND USES; TERMINOLOGY*

secondary liability is discussed in Chapter 7 Section 3F.) Normally, one indorses an instrument by signing one's name on the back of it. By indorsing, the payee bank becomes an *indorser*. The bank to whom the note is transferred is the *indorsee*. While the payee bank is in possession of the note as payee, it is the *holder*. (In the examples that appear in this book, the holder usually is identified by capital "H.") After the instrument is duly indorsed and delivered, the transferee bank becomes the new holder. (The precise meaning of "holder" is discussed on pages 103–104.)

Very often the lender insists, before making a loan, that a note be signed by someone other than the maker. If the additional party signs along with the borrower he is co-maker. More frequently, he is merely required to indorse the instrument on the back. In either case he is called an *accommodation* party if he did not receive the consideration for the instrument. The borrower, who received the money, is the *accommodated* party. (Accommodation parties are discussed on pages 164–167.) When a small or newly formed corporation wishes to borrow money, the lender usually insists that one or more of the maker's principal stockholders or officers sign as accommodation maker or indorser.

One form of note commonly used for borrowing money is the collateral note, which enables the borrower to offer as security for the repayment of the loan, not only his personal credit in the form of a negotiable note, but also some other negotiable or quasi-negotiable instrument or instruments, such as stocks, bonds, bills of lading, or warehouse receipts. The creditor creates a security interest in the collateral and perfects it by taking possession. This transaction is governed by Article 9 of the Code. When the collateral is investment securities, Article 8 will have some applicability. See 8–321 of the 1978 Edition of the Code. If the borrower's note is payable at a specified future time, as it usually is, the bank may discount it in advance or charge interest when it becomes due. An enormous amount of money is loaned on the basis of collateral notes. In some states it is common to use a note to evidence a loan made to purchase land. Such a note usually is secured by a real estate mortgage.

2. MEANS OF BUYING ON CREDIT

In a common transaction, a merchant sells goods, and in payment, receives the buyer's thirty-day note naming the seller as payee. The note is definite evidence of the debt and, as we shall see, is more satisfactory to the seller than having merely a charge on his books against the buyer. Also, the seller may indorse the note to some other firm in payment of goods he has purchased or he can raise money on the note at his bank. When the seller indorses, he normally binds himself to pay if the buyer does not. Sometimes a note passes through a number of hands before maturity, gaining credit by virtue of the secondary liability assumed by each indorser. At present, a vast amount of credit is extended to consumers of goods and services on the basis of notes, particularly installment notes which often are secured by liens on the goods purchased.

3. Means of Evidencing Previously Existing Debt

Very often a debtor who is short of funds can obtain temporary relief by giving his creditor a promissory note payable at a later time. From the creditor's point of view this is often more satisfactory than bringing suit to collect because he can maintain the good will of his debtor, who is often a customer, and secure undisputable evidence of the debt. The creditor may use the note, in turn, to pay his own debts, to sell, or to deliver as security to bolster his own credit with his bank.

B. DRAFT

Ordinary or non-check drafts, usually referred to simply as "drafts," are used extensively in business for a variety of purposes, most commonly to collect accounts, finance the movement of goods, and transfer funds.

1. Means of Collecting Accounts

Although the effectiveness of a draft as a means of collecting accounts is sometimes questioned, it is often used for this purpose. Suppose a creditor in Nebraska has sold goods to a debtor in Florida and is having difficulty in collecting. After trying other means to persuade the debtor to pay, the creditor may decide to draw a time draft on the debtor, thereby definitely putting it up to him to agree to pay by a certain time. The draft may be payable at a fixed future date or a stated time after demand or sight. The creditor (drawer) will name his bank in Nebraska as payee and will deliver the draft to this bank for collection. The creditor's bank will then forward the draft to its correspondent bank in Florida. The Florida bank will then present the draft to the debtor (drawee) and request his acceptance. "Acceptance is the drawee's signed engagement to honor the draft as presented." UCC 3–410(1). In other words, the debtor (drawee) can accept by merely signing the draft. The debtor is, of course, not obligated to accept the draft because the owing of a debt imposes no duty to assume a new form of contract for its payment. If he values his business reputation, however, he may not wish to go on record as being unwilling to pay his debts. If he chooses to accede to the demand made, he usually will write across the face of the draft, "Accepted," sign his name, and possibly indicate the date and the bank at which it is payable. (See pages 17–19.) The debtor thereby becomes an *acceptor* and the draft may be referred to as an *acceptance*. Depending upon its instructions, the correspondent bank either returns the accepted draft to the forwarding bank or holds it for collection. In the latter event, it presents the draft for payment at maturity. If paid, it remits to the forwarding bank which credits the creditor's account. (For further discussion of the part played by banks in the collection process see Chapter 13. See also Article

Ch. 2   *FORMS AND USES; TERMINOLOGY*

4 of the Code.) If the debtor accepts the draft but refuses to pay when it falls due, the creditor at least possesses evidence of the debt and has the procedural advantages that attend the holder of a negotiable instrument. See 3–307. If the debtor refuses to accept the draft, the creditor is no worse off than he was originally, since he can still sue on the original debt.

2. MEANS OF FINANCING THE MOVEMENT OF GOODS

Businessmen who ship goods constantly face the problem of obtaining payment of the purchase price as promptly as possible. One of the most common and at the same time, one of the most satisfactory, methods of financing the shipment of goods involves the use of a draft.

Assume that a merchant in Wilmington is selling goods to a merchant in Cleveland. The shipment is large and the seller is not entirely satisfied with the buyer's credit rating; or perhaps the seller must keep his funds liquid by quick collections. He therefore contracts for payment by having the buyer honor a demand draft for the price plus charges, upon delivery of a bill of lading for the goods. Accordingly, when he ships the goods, he obtains from the railroad a negotiable bill of lading representing the goods. He then draws a sight, or demand, draft for the price plus charges, naming the buyer as drawee. He delivers the draft, with the bill of lading attached, to his own bank, which he names as payee in the draft, and directs that the draft with the bill of lading attached be forwarded to the Cleveland correspondent bank for presentment to the buyer (drawee). The seller's bank complies. When the Cleveland bank receives the draft and bill of lading, it notifies the buyer. If he wishes to obtain the goods, he pays the draft and the correspondent bank surrenders the bill of lading to him. The bill of lading entitles him to require the railroad to deliver the goods to him. Meanwhile the Cleveland correspondent remits the proceeds, less its charges, to the forwarding bank which credits the seller. When papers such as a bill of lading are attached to a draft which is forwarded to the drawee, as in the above case, the draft commonly is called a *documentary* draft. (For a fuller treatment of documentary drafts, see Sections 4–501 –504 of the Code.) When no papers accompany the draft, it is called a *clean* draft. See Speidel, Summers and White, Commercial and Consumer Law 902–909 (3d ed. 1981).

A common procedure for financing the movement of goods if the seller does not consider immediate payment essential is the *trade acceptance*, which arises when a purchaser of goods agrees to pay at a *future* time by accepting a *time* draft drawn on him by the seller. The procedure is the same as when a sight draft is used, except that the buyer need not make payment before he receives the bill of lading, but he must bind himself to pay at a later time by writing his acceptance on the draft. When a trade acceptance is returned to the seller, he may hold it until maturity or indorse it and discount it at his bank, thereby obtaining ready cash for the sale. Such paper is highly regarded by banks because it has behind it the credit of the acceptor *and* the drawer.

Still another form of draft used to finance the shipment of goods is the *banker's acceptance*, which is simply a draft that has been accepted by a bank. For example, a buyer in Austin, Texas, wishes to buy goods from a merchant in Baltimore, Maryland. The buyer's credit may not be good enough to enable him to buy goods on credit outside of Austin, but his credit is good at First Bank of Austin, which agrees to finance him. The seller is willing to sell upon the well-known credit of First Bank. Accordingly he ships the goods and draws a time draft for the amount of the shipment on First Bank. He forwards the draft, with the bill of lading attached, to First Bank, which writes its acceptance across the face and returns the draft to the seller. This draft is readily salable because of the high standing of the acceptor. Meanwhile, the buyer has agreed to provide First Bank with funds to pay the draft when it falls due and First Bank has given him the bill of lading entitling him to possession of the goods. In some cases First Bank will insist on receiving some form of security from the buyer.

It will be noted that the bank acceptance accomplishes the following useful purposes: It enables sellers to make sales which they otherwise would be unwilling to make; it secures payment in a highly desirable negotiable form; it enables buyers to buy in places where their credit is unknown; and like other forms of drafts, it serves as a means of transferring funds.

C. CHECK

Serving primarily as means of payment and as vehicles for the transfer of money, checks make possible an enormous amount of business activity with a minimum amount of currency. Occasionally, a postdated check is used as a short term credit device. A check also furnishes a receipt, aids in keeping records, serves as evidence, reduces the risk of loss, destruction and theft of currency, and in other ways serves as a great commercial convenience.

Normally the issuance of a check is preceded by a contract between the bank and the drawer under which the latter makes a deposit and the bank agrees to honor his order for making payments in such amounts as he sees fit, up to the amount of deposits. (Some special problems that arise from the relationship between the bank and its checking account depositor are discussed in Chapter 13.)

A seller or other creditor who is unwilling to take his debtor's check is entitled to insist on payment in legal tender. See UCC 2–511. Such a creditor, however, may be willing to take instead a demand draft naming him as payee and drawn by the debtor's bank on the creditor's bank. If so, the debtor purchases such a draft from his bank and remits it to the creditor who obtains payment from the drawee bank which charges the drawer bank. An instrument of this kind, by which one bank draws on another, usually is referred to as a "bank draft;" but it is also a check because it is drawn on a bank and is payable on demand.

Ch. 2   *FORMS AND USES; TERMINOLOGY*

(Bank drafts are used also in the process of collecting checks deposited by bank customers, in which case they are called "remittance drafts.") The demands of a creditor who is unwilling to take his debtor's check might also be met by the debtor's purchasing from his bank, and sending to his creditor, a cashier's check, or official check, by which the debtor's bank draws on itself. Sometimes a creditor who is otherwise unwilling to take his debtor's check will take it if it has been certified by the drawee bank which becomes liable as *acceptor* of the check. (See pages 155–158.)

D. CERTIFICATE OF DEPOSIT

A firm having a temporary excess of cash may wish to invest it on a short term basis in some liquid form. An individual faced with this problem would be likely to deposit his funds in a savings account, but by law a corporation is not permitted to draw interest on savings accounts, nor on demand deposits in commercial banks. The corporation's needs may be conveniently satisfied, however, by a time certificate of deposit. Since it carries the primary obligation of a bank and bears a stated rate of interest, it is a sound investment and may serve as an effective means of making payment, as a convenient form of security, or as an asset which can be quickly liquidated if necessary.

During recent years, time certificates of deposit, usually for one or two years, have been used increasingly as a device for encouraging individuals to deposit funds in commercial banks. In return for the depositor giving up his right to withdraw his funds on short notice, the bank is willing to pay a higher rate of interest than it pays on its ordinary savings accounts.

The foregoing is, of course, only a brief and summary description of the forms and uses of commercial paper. There are many other less common uses. In fact, attorneys, bankers and other business people are constantly challenged to devise new ways to have commercial paper serve the needs of business.

This chapter has considered the special characteristics and uses of each of the various forms of commercial paper and, in doing so, has emphasized the differences among them. Therefore it seems appropriate to emphasize, before proceeding further, that most of the principles in the law of commercial paper apply equally to notes, ordinary drafts, checks, and certificates of deposit. In the discussion that follows, when this is not so, it will be clear from the context.

Although most of the instruments set forth above happen to be negotiable, the general descriptions that have been given apply to notes, drafts, checks, and certificates of deposit whether they are negotiable or not. Also, the instruments that are not negotiable may serve the same general purposes as those that are, but usually not as well. After one last brief excursion, we will consider in Chapter 3 the concept of negotiability in more detail. In Chapter 4, we will examine the special requirements that must be satisfied in order to make a note, draft or check negotiable.

Ch. 2 *FORMS AND USES; TERMINOLOGY*

§ 3. Functional Overlaps Among Articles of Code

Article 3 is but one of ten articles in the Uniform Commercial Code. The functional overlaps between and among these articles can be illustrated in a simple way. Suppose Seller, a manufacturer, contracts to sell goods to Buyer, a retailer. The formation, performance and enforcement of contracts for the sale of goods is governed by Article 2. If Seller agrees to ship the goods to Buyer, the relationship between Seller (consignor), Buyer (consignee) and Carrier under the bill of lading issued by Carrier will be governed largely by Article 7. The bill of lading, called a document of title, may or may not be negotiable. If negotiable, the form and effect are governed by Article 7, Part 5 rather than Article 3. 3–103(1). As we have just seen, if payment for goods is required at or before delivery, a combination of a negotiable bill of lading (Article 7) and a demand draft (Article 3) can be used to insure that Buyer's promise to pay (Article 2) is performed before delivery. Further, upon presentment of the demand draft, Buyer may pay either by check or cashier's check. (Article 3). Or, in some cases, the parties may have arranged for a letter of credit, whereby Seller is paid by a bank against the presentation of documents at the time the goods are shipped. (Articles 7 and 5). Finally, if payment is by check, problems arising in the process of presenting the check to and collecting from the drawee bank are governed by Article 4.

If Seller agrees to extend Buyer credit, payment is not required upon delivery. If the parties agree, Seller may create a security interest in the goods and perfect it by filing a financing statement in a public office. (Article 9). If Buyer defaults, Seller may enforce the security interest (Article 9, Part 5) and, if there is a deficiency, enforce the promise to pay for the balance. (Article 2). If Buyer has issued a promissory note, then that note will be enforced to make up the deficiency. (Article 3). At all times and in all articles, the parties are aided by the common definitions and policies contained in Article 1.

CHAPTER 3

NEGOTIABILITY AND RISK ALLOCATION IN COMMERCIAL PAPER

There are two doctrinal lynchpins in the modern law of commercial paper: (1) free transferability; and (2) holder in due course.

The doctrine of free transfer enables a person in possession of a note or draft to transfer his rights in the writing to a third person. If there is a willing buyer, the transferor is able to convert a time note or draft immediately to cash and thereby finance daily operations. The transferee's incentive to purchase the writing is the potential profit to be made on the transaction, in the interest to be paid and any discount from the face amount of the note in determining the value to be paid for it. Thus, if the transferee paid $900 for a note containing the maker's promise to pay $1,000 at 15% interest per annum, the potential annual return on the investment is 25%. The incentive to purchase, however, would be impaired if the transferee were subject to any claims or defenses that the maker or drawer could assert against the transferor. It is one thing to purchase the transferor's rights and quite another to be subject to their infirmities.

The doctrine of holder in due course evolved to insulate the transferee from these claims and defenses. The transferee must satisfy three conditions to become a holder in due course: (a) the note or draft must be negotiable, i. e., an instrument; (b) the instrument must be negotiated to the transferee so as to make him a holder; and (c) the holder must take the instrument in good faith and for value and without any notice of claims and defenses. When these conditions are met, the HDC takes free of many of the claims and defenses that a maker or drawer could assert against the payee. In short, he obtains greater protection from risk than his transferor had. See, generally, Dolan, *The U.C.C. Framework: Conveyancing Principles and Property Interests*, 59 B.U.L.Rev. 811, 812–820 (1979).

The overall conclusion has been that the benefits to be derived from a protected market for the sale of instruments outweigh the individual interest of the maker or drawer. The background of this method of risk allocation in commercial paper will be the subject of Chapter 3.

§ 1. Assignment of Contracts Rights

A. DEVELOPMENT AT COMMON LAW

Suppose that Seller sold and delivered goods to Buyer in exchange for a promise to pay $1,000 in 90 days. The promise is in writing but does not qualify as a negotiable instrument. Thereafter, Seller, in writing, transfers or assigns his contract right against Buyer to a third party for value. Seller now becomes Assignor, third party becomes Assignee and Buyer becomes Debtor. What is the legal effect of this transaction?

Ch. 3 *RISK ALLOCATION*

Until the early 18th Century, English law did not recognize the transfer of a contract right not represented by an instrument or some tangible token. The transfer in writing of a chose in action was not effective and could not be enforced by the transferee against the promisor. Through the use of fictions and the enactment of legislation, the power of Assignor freely to transfer or assign contract rights was gradually recognized. The legal effects of a present assignment were: (1) Assignor's right to performance from Debtor was extinguished and that right was transfered to Assignee; (2) Assignee could enforce the contract right against Debtor, subject to the terms and conditions of the contract for sale; and (3) Debtor's consent to the assignment was not required unless the assignment materially altered Debtor's risks or rights. The assignment, however, might not be effective if prohibited by statute or public policy or a valid clause in the contract between Assignor and Debtor. Finally, Debtor had a greater basis for objection if Assignor both assigned contract rights and delegated performance duties to Assignee without Debtor's consent. See UCC 2–210 and Restatement, Second, Contracts § 318.

When Assignee, now owner of the contract right, gave notice of the assignment to Debtor, Debtor was obligated to pay Assignee rather than Assignor $1,000 in 90 days. But as one commentator has observed, Assignee's rights are precarious indeed. See Gilmore, *The Assignee of Contract Rights and His Precarious Security*, 74 Yale L.J. 216 (1964). Unless Debtor has made an enforceable contract with Assignor not to assert claims or defenses against any assignee, Assignee takes the contract right subject to a number of risks: (1) the voidability or unenforceability of the right by Assignor against Debtor; (2) defenses or claims of Debtor against Assignor which arose before Debtor received notice of the assignment; (3) dischargeability of the right due to such events as failure of condition or changed circumstances; and (4) defenses and claims arising from Assignee's conduct. See Restatement, Second, Contracts § 336 and UCC 9–318. If, then, in the sale between Assignor and Debtor there was fraud or breach of warranty by Assignor, Assignee would, in theory, be subject to Debtor's claims and defenses.

How can these risks be managed by Assignee? There are several possibilities. One is to obtain an agreement by Debtor at the time of contracting that he will not assert any claims or defenses against a future assignee. Here Assignee is a third-party beneficiary of Debtor's agreement with Assignor. Another possibility is for Assignee to allocate the risk by agreement with Assignor. Thus, Assignee may either assess the risk, discount the face value of the writing to reflect it and take without recourse or enter an agreement that permits Assignee to recover from Assignor should claims or defenses materialize. Finally, Assignee could take the assignment without any agreement and rely upon warranties made by Assignor. Unless disclaimed, Assignor typically warrants:

(a) that he will do nothing to defeat or impair the value of the assignment and has no knowledge of any fact which would do so;

(b) that the right, as assigned, actually exists and is subject to no limitations or defenses good against the assignor other than those stated or apparent at the time of the assignment;

(c) that any writing evidencing the right which is delivered to the assignee or exhibited to him to induce him to accept the assignment is genuine and what it purports to be.

Restatement, Second, Contracts § 333.

Example 3

S, a dealer, sold B a used truck for $1,500. B paid $500 down and promised to pay the $1,000 balance in 60 days. S made an express warranty to B that the truck was in "operable" condition when in fact the brakes were defective and S knew it. S assigned the contract right to C for $800 "without recourse" and C gave B prompt notice of the assignment. The brakes failed and S refused either to repair the car or cancel the deal. B refused to pay C the $1,000 due and C brought suit as assignee of the contract right. C is subject to B's defenses and claims arising from the contract for sale. S' assignment "without recourse" was probably effective to disclaim any warranties otherwise made by assignment, although that language would not be an effective disclaimer under Article 3. (See pages 160–161.) In the absence of a disclaimer, Assignor warrants to Assignee, inter alia, that he has no knowledge of any fact that would impair the value of the assignment and that there are no defenses good against him other than those stated or apparent at the time of the assignment. See Restatement, Second, Contracts § 333(1). Both of these warranties appear to be broken in this case. C, then, would have an action over against S.

B. EFFECT OF UCC ARTICLE 9, SECURED TRANSACTIONS

The foregoing "common law" analysis must be modified in light of the enactment of UCC Article 9. In our hypothetical case, S' contract right to payment is defined as an account in Section 9–106. Article 9 applies to "any transaction (regardless of its form) which is intended to create a security interest in * * * personal property * * *," including "any sale of accounts" and security interests "created by contract including * * * assignment." See 1–201(37) (definition of security interest) and 9–102(1) and (2). The starting point is the conclusion that any assignment or "sale" of accounts by S to C in the ordinary course of business is intended to create a security interest in accounts and is within the scope of Article 9. The exclusion in Section 9–104(f) of some transactions in accounts is not applicable here.

What is the impact of Article 9 upon the assignment of accounts? First, Assignee's security interest in the account will not be enforceable against either Debtor or Assignor unless it is created in accordance with Section 9–203. At a minimum, (1) there must be a written security agreement, see 9–105(1)(l), signed by Assignor and containing a description of the accounts, (2) Assignee must give value and (3) Assignor must have rights in the collateral. Second, to insure that the security interest will have maximum priority over third parties, particularly other creditors competing for the same account, it must be perfected. See 9–303(1). Normally, a financing statement must be filed to give public notice in a designated place within the state. See 9–302(1), 9–303, 9–401 and 9–402. Filing is not required, however, when the assignment "does not alone or in conjunction with other assignments to the same assignee transfer a significant part of the outstanding accounts of the assignor." 9–302(1)(e). This exception may apply in our hypothetical so that C could have a perfected security interest without filing. Again, perfection is necessary for C to have priority over competing claims to the same account should default occur. Third, even if the security interest is created and perfected, C is not home free against B. Unless B (called an Account Debtor under Article 9) has made an enforceable agreement with S (called a Debtor under Article 9) not to assert claims and defenses against C (called a Secured Party under Article 9), Section 9–318 preserves the risk ratio to which an assignee was subject at common law. The rights of C, for example, are subject to the terms of the contract between S and B and "any defense or claim arising therefrom" and "any other defense or claim of the account debtor against the assignor which accrues before the account debtor receives notification of the assignment." 9–318(1).

The point to remember is that whether Assignee takes by assignment or sale of the account, the interest created is a security interest which is subject to Article 9. The assignee who ignores Article 9, therefore, may have an interest which is unenforceable against either Assignor or Debtor and which will not have priority over third parties who assert competing claims to the same collateral. See 9–201, 9–301 and

9–312(5). But even if Article 9 is satisfied, the assignee as Secured Party does not really improve his risk position against Debtor. Unless there is a contrary agreement, the rights of Assignee are still subject to Debtor's claims and defenses as stated in Section 9–318.

Example 4

The facts are the same as Example 3 except that C failed properly to create a security interest under 9–203. In addition, S, the day after the assignment to C, assigned the same account to D. D properly created a security interest in the account but failed to perfect it by filing. Finally, S, the day after the assignment to D, assigned the same account to E. E properly created a security interest in the account and perfected it by filing. Thereafter, the brakes failed, S failed to repair them and B refused to pay either C, D or E. On these facts: (1) C's interest is not enforceable against S or any third party, including B, 9–203(1); (2) D's interest is enforceable but as an unperfected security interest is subject to E's later but perfected security interest, 9–301(1)(a) and 9–312(5); and (3) both D and E, regardless of priority between themselves, are subject to B's claims and defenses under 9–318.

§ 2. Negotiability Under Article 3: A Contrast with Assignment

It is now time to contrast the fate of an assignee under common law or Article 9 with that of a "transferee" of an instrument under Article 3. The contrast is vital and will provide a brief overview of Article 3.

Let us continue our hypothetical employed in Examples 3 and 4. S sells B a used truck for $1,500, gets a

$500 down payment and takes a negotiable promissory note from B. The note consisted of a writing signed by B which contained an unconditional promise to pay "S or his order" the sum of $1,000 "60 days from date" with interest at 12%. See 3–104(1). There you have it—a negotiable promissory note or, more generally, an instrument under Article 3. So what? How does this change the risk analysis associated with the assignment of contract rights? The answer requires a few deft steps under Article 3 and familiarity with some new terminology.

The first step is the "issue" or first delivery of the instrument by Maker (B) to Payee (S). 3–102(1)(a). At this point, Payee becomes a Holder, i. e., a "person who is in possession of * * * an instrument * * * issued * * * to him or his order * * *." 1–201(20). As such, Payee has the right to transfer or negotiate the instrument or to "enforce payment in his own name." 3–301. At the same time, when Payee takes the instrument "for an underlying obligation," that obligation is "suspended pro tanto until the instrument is due or if it is payable on demand until its presentment." 3–802(1). Thus, upon issue Payee becomes a Holder and, for a defined period, the instrument becomes the exclusive embodiment of the underlying right. It is tangible personal property which can be transferred by physical delivery or serve as collateral in which a security interest can be created.

Suppose, now, that after 60 days Maker fails to pay upon presentment by Payee-Holder. This failure (or dishonor) breaches the contract of Maker, 3–413(1), and gives Payee the option of suing on the instrument or the underlying obligation. 3–802(1)(b). But remember, Maker will, at a minimum, defend upon the ground of breach of warranty by Payee. We know that the defense would be available if Payee sued on the underlying obligation. Suppose he sued on the instrument? Clearly, there would be some procedural advantages if he did so. See 3–307. But, alas, Payee would be subject to the defense of breach of warranty. Why? Because Payee, although a Holder, is not a holder in due course. At a minimum, he took the instrument with "notice" of a "defense against or claim to it on the part of any person." 3–302(1)(c). See also 3–305(2). He is, therefore, "subject to * * * (b) all defenses of any party which would be available in an action on a simple contract." 3–306(b). For the payee, therefore, taking an instrument interjects form and procedural advantage without changing the substance of the risks in the underlying transaction.

The second and critical step is to make some third person a "holder in due course" of the instrument. Why? Section 3–305 provides that a holder in due course "takes the instrument free from * * * (2) all defenses of any party to the instrument with whom the holder has not dealt," with exceptions not relevant here. 3–305(2). The first question is how does this third person become a holder in due course? The second question is why does Article 3 give the holder in due course greater protection than the assignee of an ordinary contract right? The answer to the first question is derived from an analysis of the statute. The answer to the second is found in the policy justifications behind the holder in due course doctrine.

To constitute the third party a holder, Payee must negotiate the instrument to him. Since the note is "order" paper, i. e., it is payable to "Payee or order," negotiation is by delivery of the instrument with any necessary indorsement. See 3–202(1) and 3–110. Thus, physical delivery with the indorsement "Pay to C, (signed) S" would be sufficient to make C a holder. See 3–204(1) and 1–201(2). But unless C is a holder in due course, he is subject to the same defenses as Payee.

Whether C is a holder in due course depends upon Section 3–302(1). C must be a holder *and* take the instrument: (1) for value, see 3–303; and (2) in good faith, see 1–201(19); and (3) without notice of any defense or claim to it on the part of any person, see 3–304. If these conditions are met, C becomes a holder in due course entitled to the special protection carved out by Section 3–305. Thus, Maker would be precluded from asserting the defense of breach of warranty against HDC. 3–305(2). If C is not a holder in due course, he is, much like the assignee of a contract right, subject to Maker's defense. See 3–306. There is, in this situation, some warranty protection available against Payee. See 3–417(2). But this does not cover the risk that Maker will have the defense of breach of warranty against Payee. More probably, C will sue Payee on the instrument to enforce Payee's contract of indorsement. See 3–414(1). Much more on all of this later.

Example 5

On July 1, S sold B goods to be delivered on August 1 for $1,000. B signed a writing promising to pay S $1,000 on August 10 with interest at 12%. On July 15, S indorsed the note and delivered it to C for value. C took it in good faith and without notice of any claim or defense. On August 1, S failed to deliver the goods to B and declared bankruptcy. On August 10, B refused to pay the note when presented by C and C brought action to recover. C is subject to B's defense of "failure of consideration." 3–306(c). The note was not negotiable because it was not payable to "S or order." It was not, therefore, an instrument and C could not be a holder even though the "due course" conditions were satisfied. The analysis but not the result is altered by Section 3–805. Since the sole defect is that the note was not payable "to order," the note is subject to Article 3 but C cannot be a holder in due course. Thus, the scope of C's risk is determined by Section 3–306 rather than the common law. Similarly, C's rights, if any, against S are determined by Article 3.

§ 3. Why Negotiability?

What explains the special protection afforded to the holder in due course of an instrument by commercial law? Part of the answer may be historical. At common law, the bona fide purchaser of tangible personal property took free of equities and defenses while the purchaser of a chose in action, did not. Once it was decided that the underlying contract right merged with the instrument upon issue, it was but a short step to the conclusion that the good faith purchasers of this form of tangible personal property also "took free." A more complete answer, however, must take into account perceived commercial needs for a protected market in which "bills and notes" could be freely and safely traded. The requirements of form on the face of the writing and in the process of negotiation, the economic advantages of free transferability and the reliance by purchasers "in due course" on that form as new value was given in the various credit markets all reflect a body of law developed to facilitate trade in these "specialties" or "couriers without luggage."

Recently, some commentators have questioned whether the concept of negotiability is still needed. One argument is that the concept produces an allocation of risk that is unfair to Makers and Drawer, in that they may not be adequately informed of the consequences of signing an instrument. Another argument is that the current scheme is inefficient. Why put the initial risk on Maker and force an action over against Payee (who may be insolvent or difficult to sue) when Holder may be in the best position to assess and provide for the risk at the time the instrument is taken? If legitimate commercial needs, including risk allocation, can be met by more efficient devices, why put the initial and perhaps ultimate burden on Maker?

These arguments have been quite persuasive when Maker is a consumer who is buying or borrowing for personal, family or household purposes. We will examine some of the "anti-negotiability" cases and legislation in Chapter 10, Section 14. In commercial transactions, however, the concept of negotiability and the holder in due course still prevail. Assuming for now that this concept has continuing vitality as a risk allocation device, we will now consider in more detail the requisites of negotiability.

CHAPTER 4

REQUISITES OF NEGOTIABILITY

To be an instrument under Article 3, a writing must possess the requisites set forth in UCC 3–104(1). If one requisite is missing, it is not negotiable. The writing may evidence a valid contract and be assignable from person to person; but with few exceptions, it is subject to the disabilities of assignment discussed in Chapter 3, Section 1. On the other hand, if the writing possesses all of the requisites of negotiability, it may, after issue to the first holder and negotiation to a holder in due course, entitle that holder to the special protection afforded by UCC 3–305. Thus, it is important to have a firm understanding of the requisites for negotiability.

§ 1. Requisites of Negotiability Under Article 3: 3–104(1)

The needs of the merchants who first used commercial paper required clear and concise language, which could be quickly recognized and easily understood and which would facilitate the transfer of such paper in the course of trade. In the contemporary world, the need for concise language in notes, checks and drafts is increased because of the velocity of their circulation. Prospective purchasers must be able quickly to understand and evaluate a writing from what appears on its face.

Ch. 4 REQUISITES OF NEGOTIABILITY

These basic needs of commerce are recognized by the Code just as they were recognized by common law and under the N.I.L. Section 3–104(1) states the requisites of negotiability as follows:

Any writing to be a negotiable instrument within this Article must

(a) be signed by the maker or drawer; and

(b) contain an unconditional promise or order to pay a sum certain in money and no other promise, order, obligation or power given by the maker or drawer except as authorized by this Article; and

(c) be payable on demand or at a definite time; and

(d) be payable to order or to bearer.

Assuming that a writing is signed by a maker or drawer, the answer to the question whether it satisfies the requirements for negotiability in Section 3–104(1) must be found on the face of the writing. Comment 5, 3–119. If for example, the promise on the face of the writing is simply to "pay Payee," the presence on the back of the writing or in a separate agreement of the magic words "or order" will not be sufficient. On the other hand, if a writing is negotiable on its face the presence of a separate agreement, oral or written, "does not affect the negotiability of an instrument." 3–119(2). See 3–105(2)(a) and Booker v. Everhart, 294 N.C. 146, 240 S.E.2d 360 (1978). Thus, with the writing in hand and Section 3–104(1) in mind, it is usually possible quickly to determine whether the writing is negotiable. Since many consequences hinge upon negotiability, this Section is the logical starting place for study. Even so, there will always be some writings which cannot be categorized as negotiable or not solely on the basis of the broad language of Section 3–104(1). A preliminary comment on methodology, therefore, seems to be in order.

The basic requirements of negotiability in Section 3–104(1) are supplemented and amplified in Sections 3–105 through 3–111 and in Sections 3–112 and 3–118. For example, the requirement that a promise be "unconditional" in Section 3–104(1)(b) is amplified in Section 3–105 and supplemented by Section 3–112(1)(b) and the rules of construction in Section 3–118. A complete answer to any question of negotiability, therefore, requires that the entire statutory scheme be understood and applied in the light of the underlying commercial policies.

Upon close examination, it appears that there are nine basic requisites which must be satisfied if a writing is to qualify as a negotiable instrument. They will be discussed separately below.

A. MUST BE IN WRITING: 3–104(1)

The need for a writing rarely presents any problem. Although commercial paper usually is executed on printed forms drafted by lawyers with the requisites of negotiability in mind, the requirement of a writing may be satisfied also by engraving, stamping, lithographing, photographing, typing, longhand writing in pencil or ink, any similar process, or any combination of these. See 1–201(46). It is reported that a bank in Virginia once paid a check written on a coconut.

Ch. 4 *REQUISITES OF NEGOTIABILITY*

B. MUST BE SIGNED BY MAKER OR DRAWER: 3–104(1)(a)

This requirement is equally easy to satisfy. Normally it is met by a person writing his name in longhand in ink. It may also be satisfied by adopting any symbol affixed to the instrument by hand, machine or in any other manner, if it is done with the present intention of authenticating the instrument. 1–201(30). The symbol may consist of a name, initials, trade name, assumed name, mark, or even a thumb print. 3–401 and Comment 2. Although there is no clear limit to the ways in which a signature may be affixed, anyone presented with an instrument signed in some unusual way should remember that whenever the presumptive validity of a signature is overcome, the burden of establishing its effectiveness is on the person who claims under it. 3–307(1)(a). Also, an unusual signature may decrease the marketability of an instrument by creating uncertainty and requiring a collateral investigation into authenticity.

C. MUST CONTAIN A PROMISE OR ORDER: 3–104(1)(b)

To be negotiable, a note or certificate of deposit must contain a *promise* to pay, whereas a draft or check must contain an *order* to pay.

1. PROMISE

Notes usually contain an *express promise* such as "I promise to pay," but any language supporting an undertaking to pay is sufficient. Mere acknowledgment

[*52*]

of a debt, however, is not adequate. 3–102(1)(c). Accordingly such statements as "I.O.U. $100" and "Borrowed $100" have been held not to be promises. But it has been held that if such words as "to be paid on demand" or "due on demand" are added, the need for a promise is satisfied. The fact that such cases have reached the courts, however, is good reason for avoiding doubtful language. A *certificate of deposit* is different; here the requirement of a promise is satisfied by the fact that the bank's acknowledgment of the deposit and the other terms of the instrument clearly *imply* "an engagement" to pay. 3–104(2)(c). See 3–413(1), where the contract of a maker is set forth.

2. ORDER

An order is a direction to pay. 3–102(1)(b). The usual way of expressing the order is to use the imperative form of the verb "pay." "Pay bearer" is clearly an order. The mandatory character of the order is not altered by the fact that it is couched in courteous form such as "please pay" or "kindly pay." However, there must be more than an authorization or request. 3–102(1)(b). Such uncertain language as "I wish you would pay" would not qualify as an order. The commercial world wants no uncertainty about the right to be paid. Moreover, to operate as an effective order, the direction to pay must identify with reasonable certainty the person who is ordered to pay. 3–102(1)(b). (The person ordered to pay is the *drawee*.) In an ordinary draft, this requirement is usually satisfied by inserting the drawee's name immediately following the word "To" on the printed form. (See the form draft on page 18.) In the case of an ordinary check it is satisfied by having the drawee bank's name printed on the face of the check.

Note that the drawer does not make an express promise to pay the payee. But he does "engage" that if certain conditions are satisfied, he "will pay the amount of the draft to the holder." 3–413(2). Thus, the drawer of a draft dishonored by the drawee may be liable on the instrument even though it does not contain an express promise to pay.

a. Drawer Named as Drawee

Occasionally, the drawer orders himself to pay; that is, the drawer and drawee are the same person. Typically this happens when an agent in one department or branch of a corporation draws on the corporation and addresses the order to another department or branch. For example, an adjuster acting for a claims department of a corporation may pay a claim by issuing a draft naming the corporation as drawee and addressing it to the treasurer's office. When the drawer and drawee are the same person, the instrument is effective as a promissory note. 3–118(a). Consequently, the drawer's liability is the same as normally is imposed on the maker of a note rather than that normally imposed on a drawer. This means that there is no need for presentment, dishonor and notice of dishonor which normally are required to charge the drawer as explained in Chapter 8.

b. Multiple Drawees

Although an order must identify the drawee with reasonable certainty, it may be addressed to more than one drawee, either *jointly* (To A *and* B) or in the *alternative* (To A *or* B). 3–102(1)(b). In either case the holder is required to present the draft only once. If the instrument is not paid or accepted as drawn when it is presented to any one of the drawees named, it is dishonored, and the holder may proceed immediately against the drawer without presenting it to the remaining drawees. 3–504(3)(a) and Comment 3. This enables a large corporation to issue a negotiable dividend check naming a number of independent drawees in the alternative in different parts of the country. This is a convenience to the stockholder who can present the check to any drawee, and if that drawee dishonors it, he can proceed immediately against the drawer. Naming drawees in *succession* (To A, and if A fails to pay, to B) prevents negotiability because in this case a holder has no right to proceed against the drawer until the draft has been dishonored by all drawees. 3–102(1)(b) and Comment 3.

D. PROMISE OR ORDER MUST BE UNCONDITIONAL: 3–104(1)(b); 3–105

Two major functions of commercial paper are to serve as a substitute for money and as a reliable basis for credit. If commercial paper is to perform either function effectively, businessmen must be assured that there are no strings attached to payment. Therefore, an instrument is not negotiable unless it contains

Ch. 4   REQUISITES OF NEGOTIABILITY

an unconditional promise or order. 3–104(1)(b). See 3–105. That is, the obligation must be expressed in terms which are absolute and not subject to contingencies, provisos, qualifications, or reservations which may impair the obligation to pay. It must be a "courier without luggage." Overton v. Tyler, 3 Pa. 346, 347 (1846). Thus, a writing which provided that it could not be pledged, assigned or transferred without the maker's written consent was held to contain a conditional promise to pay. This precluded negotiability. First State Bank at Gallup v. Clark, 91 N.M. 117, 570 P.2d 1144 (1977).

1. TERMS OF INSTRUMENT DETERMINE PRESENCE OF CONDITION

As far as negotiability is concerned, determining if a promise or order is conditional *requires an examination of the instrument itself and nothing else.* No matter what anyone has said about the instrument, for the purpose of determining negotiability the promise or order it contains is unconditional unless something in the instrument itself expresses the contrary. For example, assume that M hands P a note which provides, "I promise to pay $1000 to the order of P on demand. (Signed) M," and while doing so, M says, "This money will not be paid unless a bond is delivered to me." The note is negotiable because *the promise contained in the note itself* is unconditional and all other requirements are satisfied. The same is true if the condition is expressed in a separate writing prepared contemporaneously with the note. See 3–119(1). (See Chapter 5, Section 4B, for the effect of the extrinsic condition.) In contrast, if the note *itself* provides that payment is to be made only if the bond is delivered, the note is not negotiable. 3–105(2)(a).

2. EFFECT OF ANOTHER WRITING: 3–119

Negotiable instruments are rarely issued in isolation. Usually they are issued pursuant to or as part of some underlying agreement between the parties. For example, a check may be delivered to carry out the terms of a sales contract executed some time earlier, or a note may be issued according to the terms of a loan agreement which provides for other notes as well as for security.

When determining the respective rights and duties of the *original* parties to transactions of this kind, the Code follows the common law rule that writings executed as part of the same transaction are to be read together as a single agreement. <u>Negotiability, however, is concerned primarily with the rights of those who are not parties to the original transaction.</u> Consequently, *<u>regardless of what any outside writing may provide, it does not destroy the negotiability of an instrument which otherwise satisfies the requirements of negotiability</u>*. 3–119(2). When the instrument itself makes express reference to an outside agreement or document, however, the effect on negotiability may vary considerably from case to case depending upon the *nature of the reference*. A statement that the note "arises" out of a separate agreement does not make a promise conditional, 3–105(1)(c), but a statement that the promise was "dependent" upon the terms of that agreement or "incorporates them by reference" precludes negotiability. See Booker v. Everhart, 294 N.C. 146, 240 S.E.2d 360 (1978), where the court stressed that the critical question was whether all of the essential terms could be determined from the face of the instrument. If terms in another writing, whether conditional or not, are incorporated by reference in the note, the answer is no.

3. IMPLIED CONDITIONS NOT RECOGNIZED

In treating such references, the Code strongly favors negotiability by providing that a promise or order which is otherwise unconditional is not rendered conditional by the fact that the instrument is subject to implied or constructive conditions. 3–105(1)(a). The Code thus rejects the theory that a recital in an instrument disclosing that it is given for an executory promise creates an implied condition that the instrument is not to be paid unless the promise is performed and so destroys negotiability. Commercial paper often refers to contractual arrangements still to be performed. Usually such recitals are intended as statements of consideration without implications of any kind. If the maker or drawer intends to condition his liability, the Code requires him to do more than leave it to inference. Thus, if an instrument otherwise negotiable provides that it is given as payment in accordance with or "as per" some collateral contract, the possible inference that payment is conditioned on performance of the other contract is not permitted to destroy negotiability. 3–105(1)(b). Rather it is treated merely as a recital of the origin of the instrument or as an infor-

REQUISITES OF NEGOTIABILITY Ch. 4

mational reference to another contract. In like vein, a statement in a draft that it is drawn under a letter of credit does not state a condition but merely identifies the occasion for the issuance of the instrument. 3–105(1)(d). Also, the statement in an instrument that it is secured by a mortgage or other security device does not render it conditional. 3–105(1)(e). The existence of the security, rather than detracting from the obligation, fortifies it. See also, 3–112(1)(b).

4. "SUBJECT TO" OR SIMILAR WORDS NEGATE NEGOTIABILITY

If the instrument expressly states that it is "subject to" or "governed by" any other agreement, or contains any other words which, fairly construed, convey the same meaning, the promise or order is conditional and negotiability is destroyed. 3–105(2)(a) and Comment 8. Here the limitation appears in the instrument itself and no inference is necessary. The fact that the terms of payment cannot be determined by looking at the instrument itself, that it is necessary to look to an outside agreement, is contrary to the concept of negotiability. This is so even though it appears by hindsight that the outside document, to which the instrument is expressly made subject, contains no conditions or other provisions that are contrary to the requirements of negotiability. Booker v. Everhart, 294 N.C. 146, 240 S.E.2d 360 (1978).

If, however, the instrument unconditionally provides terms of payment to which the holder is entitled, the promise or order is not prevented from being uncondi-

Ch. 4   *REQUISITES OF NEGOTIABILITY*

tional by the fact that it also declares that a right of *prepayment* or *acceleration* is provided in a separate writing. 3–105(1)(c). In this case the reference can only speed up the right to payment provided in the instrument and cannot provide any impediment to the right. In substance, the reference operates much like an acceleration clause which, as will be explained, does not destroy negotiability by rendering the instrument indefinite as to time.

5. REFERENCE TO ACCOUNT OR FUND

The fact that an instrument indicates a particular account to be debited or any fund or source from which reimbursement is *expected* does not render a promise or order conditional. 3–105(1)(f). For example, the drawer of a bill may direct the drawee to pay the money to the order of the payee and "charge the same to the account of the drawer" or to the "merchandise account" or the like. Such directions are for accounting purposes, and the drawer's obligation is in no way contingent upon a credit balance in the account.

It is the general rule, however, that a promise or order is conditional if the instrument states that payment is to be made *only* out of a particular fund or source. 3–105(2)(b). Words of explicit limitation are required. See Bank of Viola v. Nestrick, 72 Ill.App.3d 276, 49 Ill.Dec. 309, 390 N.E.2d 636 (1979). In that case, the obligation to pay is necessarily contingent upon the sufficiency of the fund. To be unconditional an order or promise must carry the *general credit* of the maker or drawer.

[*60*]

The Code recognizes two exceptions to the rule that a promise or order to pay only out of a particular fund destroys negotiability. The first is recognized if an instrument is issued by a government agency or unit. 3–105(1)(g). Government agencies often find it necessary to issue commercial paper which provides for payment only out of special assessments of the property benefited. By rendering such instruments negotiable, it is possible to avoid disappointing purchasers who normally consider such paper to be negotiable. The second exception is recognized if an instrument is issued by a partnership, unincorporated association, trust, or other estate, and provides that payment is to be made only out of the assets of the issuing party. 3–105(1)(h). So far as negotiability is concerned such entities are treated as if they were corporations. It should be recognized, however, that this exception relates only to negotiability. It does not change any state law which declares void any provision limiting the liability of a partner, association member, trustee, or estate representative. Comment 7, 3–105.

Example 6

In a written contract dated May 15, S sold B a painting for $50,000 and agreed to have it restored before delivery on December 1. B issued a note to S dated June 1 which contained a promise to pay "S or order" the sum of $50,000 from the "assets of the B family trust" on December 1. The note also contained the following: "Reference is made to a separate contract for sale, dated May 15, whose terms are incorporated by reference." The note is not negotiable because the holder cannot determine all essential terms from the face of the writing. Comment 8, 3–105. A separate

Ch. 4   *REQUISITES OF NEGOTIABILITY*

writing must be consulted. Furthermore, the phrase "incorporated by reference" may be interpreted as subjecting the note to the terms of the separate writing. 3–105(2). The reference to a separate writing or the indication of a fund from which payment will be made, however, would not impair negotiability. See 3–105(1)(e), (f) and (g).

E. MUST SPECIFY A SUM CERTAIN: 3–104(1)(b); 3–106

As a general rule, the term *sum certain* means a sum which the holder can determine from the instrument by any necessary computation at the time the instrument is payable, without reference to any outside source.

For example, an instrument which contains a promise to pay a stated sum on demand with interest at six per cent provides for a sum certain because, even though the amount to be paid is not known when the note is issued, it can be computed readily by the holder by examining the instrument when a demand for payment is made. 3–106(1)(a). Similarly, the sum payable is certain if it is subject to a fixed discount or increase if paid before or after the date provided for payment. 3–106(1)(c). Also, the sum is certain even though it is to be paid with different stated rates of interest before and after default or on a specified date. 3–106(1)(b). And the fact that an instrument makes no reference to interest does not make the sum uncertain since such an instrument simply draws no interest. Nor is the sum rendered uncertain by the fact that the instrument provides for payment "with interest," without mentioning the rate, for in that case the rate is the same as the judgment rate which is declared by law. 3–118(d). It seems to be assumed that everyone knows this rate. The sum is not certain, however, if the instrument is payable with interest "at the *current* rate" because the current rate of interest cannot be determined from the instrument itself without reference to an outside source. Comment 1, 3–106.

Commercial needs sometimes require some relaxation of the rule that a *sum certain* is a sum which the holder can determine from the instrument itself with any necessary computation at the time the instrument is payable. For example, promissory notes often provide for collection charges to make certain the holder will receive his principal sum without impairment due to collection costs. The Code states that if an instrument provides for payment with costs of collection or an attorney's fee or both on default, the sum payable is a sum certain. 3–106(1)(e). If the cost or rate is not stated, it usually is supplied by local law or custom. Where the instrument itself expressly provides the amount of the attorney's fee, the amount must be reasonable or the provision will be held to be a penalty and void because it is against public policy. The Code relaxes the general rule also by providing that negotiability is not destroyed by making the instrument payable with, or less, *exchange* at the *current* rate 3–106(1)(d), "exchange" being the difference between the values of the two currencies which is determined by supply and demand and fluctuates from day to day

Ch. 4   *REQUISITES OF NEGOTIABILITY*

so that it can be known only by looking beyond the instrument.

F. MUST BE PAYABLE IN MONEY: 3–104(1)(b); 3–107

An instrument is payable in money if it is stated as payable in a medium of exchange authorized or adopted by a domestic or foreign government as part of its official currency at the time the instrument is issued. 1–201(24); 3–107(1); and Comment 1, 3–107. The Code rejects the view of some early cases that money is limited to legal tender which an obligee is required to accept in discharge of his obligation. Comment 1, 3–107. It also rejects the contention that a commodity used as money, such as gold dust or beaver pelts, is money. Comment 1, 3–107. An instrument which is payable in "currency" or "current funds," however, is payable in money. 3–107(1). If an instrument is payable in sterling, francs, lira, or other recognized currency of a foreign government, it satisfies the requirement of money even though it is payable in the United States. Comment 1, 3–107. However, such an instrument normally may be satisfied by the payment of dollars at the appropriate rate.

Example 7

Maker signed a note, otherwise negotiable, promising to pay Payee or his order 50,000 Swiss francs "six months from date." The note was to be paid in New York City. At the date of the note, $1 could be exchanged for 1.75 Swiss francs. Payee negotiated the note to Holder for value, in good faith and without notice. On the date for payment, Holder presented the note to Maker in New York and demanded payment in U. S. dollars. At that time, $1 could be exchanged for 1.5 Swiss francs. Maker, who had defenses against Payee, argued that the note was not negotiable, first because Holder had demanded U. S. rather than Swiss "money" and, second because the amount of U. S. money, in any event, could not be determined with certainty from the face of the instrument. Maker loses on both counts. A promise to pay 50,000 Swiss francs is payable in money, 3–107(1), and unless francs are specified as the "medium" of payment, Holder may demand payment in either francs or dollars. 3–107(2). Also, the promise to pay Swiss francs "is for a sum certain in money" even though the exchange rate to dollars cannot on the date of payment be determined from the face of the instrument. 3–107(2) and Comment 4. The risk of currency fluctuation can be avoided if Maker promises to pay in Swiss francs and the note limits the medium of exchange to that currency.

G. MUST BE PAYABLE ON DEMAND OR AT A DEFINITE TIME: 3–104(1)(c); 3–108; 3–109

The basis for this requirement is that the time of payment of commercial paper is important for all parties concerned and uncertainty in this matter is commercially objectionable.

1. PAYABLE ON DEMAND

An instrument is payable on demand if it contains a promise or order to pay "on demand," "at sight," or "on presentation," or *if no time for payment is stated.* 3–108. Most instruments that are payable on demand, including virtually all checks, are so payable because they make no express provision for the time of payment.

2. PAYABLE AT A DEFINITE TIME

An instrument is payable at a definite time if by its terms it is payable on or before a stated date or at a fixed period after a stated date. 3–109(1)(a). For example, an instrument would be payable at a definite time if it were payable "on July 1, 1982", "thirty days after date" (assuming, of course, the instrument is dated), or "on or before March 1, 1982." Although "on or before" may suggest indefiniteness, there is no reason to deny such instruments negotiability because there is no more uncertainty involved in such an instrument than in an instrument which is payable on demand. Comment 4, 3–109. Similarly, an instrument payable at a fixed period after sight is payable at a definite time. 3–109(1)(b). The holder controls the matter and may, by presenting it for acceptance, have maturity promptly determined.

3. ACCELERATION CLAUSES

If the time provided for payment is otherwise definite, the fact that the instrument provides that the time of payment is subject to *acceleration* does not render the time indefinite. 3–109(1)(c). This is so whether the acceleration is at the option of one of the parties or is automatic upon the occurrence of some event. Comment 4, 3–109. But see 1–208, imposing some limitations upon the exercise of a reserved power to accelerate at will. See also 3–105(1)(c).

4. EXTENSION CLAUSES

Also, an instrument is payable at a definite time if, by its terms, it is payable at a definite time subject to (1) extension at the option of the *holder*, or (2) extension to a further definite time at the option of the *maker or acceptor*, or (3) extension to a further definite time automatically upon or after a specified act or event. 3–109(1)(d).

As indicated above, where the extension is at the option of the *holder*, the time is definite even though no time limit is stated for the extension. Unless otherwise specified, however, a consent to an extension authorizes a single extension for not longer than the original time period. 3–118(f). Furthermore, a holder may not exercise his option to extend the time over the objection of the maker or acceptor or other party who duly tenders full payment when the instrument is due. 3–118(f). This prevents a holder from refusing payment or keeping interest running against the wishes of a maker or acceptor or another party even though the instrument expressly states that the holder has the option to extend the time for payment. Comment 7, 3–118.

A provision allowing the *maker* or *acceptor* to extend for a further definite period does not interfere with negotiability because it is as if the instrument were payable at the ultimate date with the possibility of acceleration. Comment 5, 3–109. If it is expressly provided that there is no limit on the time the maker or acceptor may extend, however, the instrument is payable neither on demand nor at a definite time and

Ch. 4   *REQUISITES OF NEGOTIABILITY*

so is not negotiable. But if the instrument provides simply that the maker or acceptor has a right to extend the time of payment without stating a time, the time is definite because, unless otherwise specified, consent to an extension authorizes a single extension for not longer than the original period. 3–118(f).

5. EVENT CERTAIN TO HAPPEN

Before the Code, an instrument payable at the death of a named individual was payable at a definite time. Notes so payable are often issued by persons who anticipate their inheritances. Such notes are not commercially important. The Code renders these and similar instruments not negotiable by providing that an "instrument which by its terms is otherwise payable *only* upon an act or event uncertain as to time of occurrence is not payable at a definite time even though the act or event has occurred." 3–109(2). (Emphasis added.) This change eliminates one inroad into the concept of definiteness. Nonetheless it is still possible to accomplish the usual purpose of such instruments without destroying negotiability by specifying a date in the distant future when the instrument is payable and then adding that the time of payment will be accelerated by death.

Example 8

Maker signed a note promising to pay "Payee or order" $50,000 "when the Chicago Cubs win the National League pennant." The note is not payable "at a definite time." An optimist would conclude that the payment event is "uncertain as to the time of occurrence" and the cynic would conclude that it is uncertain as to the "fact" of occurrence. From either perspective, the note is not negotiable even though a miracle occurred after the note was issued. 3–109(2). On the other hand, a note payable upon November 1 "subject to acceleration" or "subject to automatic extension to December 25" when the Cubs win the pennant is negotiable. The uncertainty of an event which accelerates or extends an otherwise definite time for payment is not fatal. 3–109(1)(c) and (d). Finally, a note payable on November 1 "if" the Cubs win the pennant is conditioned upon the occurrence of an uncertain event. If the Cubs do not win, the promise is not enforceable. The note, therefore, is not negotiable because the promise was conditional rather than that the time for payment was uncertain. See 3–105(2).

H. MUST BE PAYABLE TO ORDER OR TO BEARER: 3–104(1)(d); 3–110; 3–111

To be negotiable an instrument must be payable *to order* or *to bearer*. 3–104(1)(d). Words which satisfy this requirement are called *words of negotiability*. The phrases "to order" and "to bearer" are the most common earmarks of negotiability and they most clearly indicate the intention of the maker or drawer to issue an instrument which is negotiable and subject to all the incidents which attach to this form of contract. They are not the only words that can satisfy the requirement, but for other words to be sufficient, they must be clear or recognized equivalents to these phrases. Comment 5, 3–104. (For the legal significance of the fact that an instrument is payable to order rather than to bearer, see Chapter 6.)

Ch. 4 *REQUISITES OF NEGOTIABILITY*

1. Payable to Order

An instrument is payable to order when "by its terms it is payable to the order or assigns of any person therein specified with reasonable certainty, or to him or his order, or when it is conspicuously designated on its face as 'exchange' or the like and names a payee." 3–110(1). The usual expressions are "to the order of X" or "to X or his order." These magic words inform the prospective purchaser that the writing is negotiable and alert the maker or drawer that if the instrument is negotiated to a holder in due course any claims and defenses assertable against the payee will be "cut off."

An "order" is a direction to pay which must identify the person to pay with reasonable certainty. 3–102(1)(b). Thus, when the maker of a note promises to pay a named payee "or order" he has empowered the payee to direct him, through the process of negotiation, to pay a designated third party or bearer. See 3–102(1)(b) and 3–413(1). The drawer of a draft, however, orders a named drawee to pay the designated payee "or order." The draft, therefore, involves a double order, first, that of the drawer to drawee to pay payee "or order," and second, that of the payee to drawee to pay a named third party or bearer if the draft is negotiated. If the drawee fails to pay as ordered by either the drawer or the payee, the drawer becomes liable on the instrument to the holder. 3–413(2).

The phrase "or assigns" seems merely to be tolerated as an intended equivalent of "or order." The reason an instrument conspicuously designated "exchange" and naming a payee is recognized as being payable to order is that it is so recognized by the Geneva Convention on international usage and is likely to be so understood by businessmen using such instruments. New York Law Revision Commission Report Legislative Document No. 65D (1955) 51.

An instrument may be made payable to the order of anyone, including the maker, drawer, or drawee. 3–110(1)(a),(b). An instrument may be made payable to the order of two or more persons together as "A, B *and* C," or in the alternative as "A, B, *or* C." 3–110(1)(d). Multiple *payees* should not be confused with multiple *drawees*. (For a discussion of multiple drawees see page 55. For further discussion of multiple payees see pages 119–120.)

The Code provides that an instrument may be made payable to the order of an estate, trust, or fund, without regard to whether or not it is a legal entity. 3–110(1)(e). Such instruments are held to be payable to the estate representative. He, of course, must account to the estate. Similarly, it may be made payable to the order of a partnership or unincorporated association, for example, "Local 10, United Plumbers." In this case it may be indorsed or transferred by any duly authorized person. 3–110(1)(g). Also, an instrument may be payable to the order of an office or officer by his title as such, for example, the County Treasurer. If so, it is payable to the principal, but the incumbent of the office or his successor may act as if he were the holder. 3–110(1)(f).

Ch. 4 *REQUISITES OF NEGOTIABILITY*

2. Payable to Bearer

An instrument is payable to bearer when by its terms it is payable "to bearer," "to the order of bearer," to a specified person "or bearer," "to cash," "to the order of cash," or to any other designation which does not purport to refer to a specific payee. 3–111. Perhaps the most common of these is "to bearer." "To the order of bearer" or "to the order of cash" or similar language is usually the result of inserting "bearer" or "cash" in the space provided for a payee's name in a check or other printed form originally containing the words "to the order of." Comment 1, 3–111. Typical designations, in addition to "cash," which do not purport to refer to specific payees are "cash for payroll," "sundries," "accounts payable," "petty cash," and "bills payable." Where an instrument contains the words "to the order of" followed by blank space, the instrument is not payable to bearer; it is an incomplete order instrument. Comment 2, 3–111. Although, as indicated above, an instrument that is payable to "John Jones, or bearer," is payable to bearer, an instrument payable to "John Jones, bearer," is not, and is therefore not negotiable.

Sometimes a person signs an instrument in which the name of the payee appears between "Pay to the order of" and "or bearer." Whether such an instrument is held to be an order instrument or a bearer instrument depends upon additional facts. If all of the words except the name of the payee are printed on a form (which is usually the root of the problem) the instrument is payable to order on the assumption that the party issuing it overlooked the words "or bearer" and that his intention was probably to sign an order instrument. If the word "bearer" is either handwritten or typewritten this is deemed to be a sufficient indication of a contrary intention and so it is held to be a bearer instrument. 3–110(3) and Comment 6, 3–118(b). Of course, in determining whether the instrument is negotiable it makes no difference whether it is payable to order or to bearer so long as it is payable to one or the other. It is necessary to classify it as order or bearer paper, however, to determine what is required for further negotiation. For now, it suffices to know that "bearer" paper is negotiated by delivery alone and that "order" paper is negotiated by "delivery with any necessary indorsement." 3–202(1).

I. MUST CONTAIN NO OTHER PROMISE, ORDER, OBLIGATION OR POWER

Even assuming that a writing fully satisfies all of the requirements of negotiability, it is prevented from being an instrument if, in addition to the promise or order to pay money, the maker or drawer gives any other "promise, order, obligation or power * * * *except as authorized by (Article 3)."* 3–104(1)(b). (Emphasis added.) For example, an instrument is not negotiable if, in addition to the promise or order to pay money, it contains a promise or order to render services or to sell goods or gives an option to acquire property or services. The principal exceptions authorized by Article 3 are expressly stated in Section 3–112, which provides that negotiability is not destroyed by giving the holder power to realize on or dispose of col-

Ch. 4   *REQUISITES OF NEGOTIABILITY*

lateral, or by giving the power to confess judgment on the instrument if it is not paid when due, or by assuming an obligation to waive the benefit of any law intended for the benefit of the obligor. Other exceptions are provided by Section 3–105 which authorizes a provision that empowers or orders a drawee to debit a particular account; by Section 3–106 which authorizes the inclusion of a promise to pay the cost of collection or attorney's fees upon default; and by Section 3–109 which authorizes a provision giving the holder an option to extend or accelerate the time of payment. It should be noticed that each of these exceptions is intended to strengthen the promise or order to pay money and has no independent value of its own.

§ 2. Instrument Declared to Be Negotiable

After tediously examining the requirements for negotiability, one might ask whether the parties can shortcut the matter by simply declaring on the writing that it is "intended to be negotiable?" Put another way, to what extent can the maker and payee vary by agreement the requirements of Section 3–104(1)?

Prior to the Code the courts were divided, (see Foley v. Hardy, 122 Kan. 616, 253 P. 238, 50 A.L.R. 422 (1927)). Article 3 does not deal directly with the question. But Section 1–102(3) states that the effect of Code provisions may be varied by agreement "except as otherwise provided in this act." Arguably, 3–104(1) is such a limiting provision. Thus, a statement on a writing payable in other than money or containing a conditional promise that "This Note is Negotiable" would not, in all probability, be effective. See 1–102,

Comment 2. But if an otherwise negotiable note lacks the magic words of "payable to order," the defect might be overcome by placing the word "negotiable" on the face of the instrument. The agreed language is intended as a substitute for the magic words and does not undercut the commercial justifications for the concept of negotiability. This result is consistent with Section 3–110(1) which declares that an instrument is payable to order if it is payable to "assigns" of a specified person or if it is conspicuously designated on its face as "exchange," or the like, and names a payee. If the parties select different language which clearly indicates that they intend the consequences associated with negotiability, the conclusion that the writing is an instrument seems justified. See also Comments 2 and 5, 3–104 and Comment 3, 3–112.

Even if the writing is not an instrument, the maker and payee may agree that the maker will not assert claims and defenses against third parties who purchase the writing for value, in good faith and without notice. The third parties are beneficiaries of the agreement not to assert claims and defenses. Compare 9–206(1). Consistent with the trend against negotiability in consumer transactions, however, many courts and legislatures have declared that such agreements in consumer contracts are against public policy. See Chapter 10, Section 14(A).

§ 3. Instrument Lacking Only Words of Negotiability

Special treatment is accorded to an instrument that satisfies all of the requirements of negotiability except that it lacks words of negotiability, in that it is not payable to either order or bearer. Such an instrument is governed by all of the special principles that govern negotiable instruments under Article 3 except that there can be no holder in due course of such an instrument. 3–805. In contrast, instruments that fail to satisfy any of the other requirements of negotiability are treated in most respects as simple contracts. Comment, 3–805. Although some courts commonly refer to an instrument that is not negotiable for any reason as a "non-negotiable instrument", other courts reserve the term "non-negotiable instrument" to refer exclusively to an instrument which is not negotiable only because it is not payable to order or bearer. Usually, the context clearly indicates the sense in which the term is used.

Example 9

Maker signed a writing dated July 1 promising to pay $10,000 to either Peter or Paul "six months after date." The note was issued to Peter. At maturity, Peter, without the knowledge of Paul, presented the note to Maker and was paid in full. Paul argued that the writing was not negotiable because it was not payable to the "order" of either Peter or Paul. Article 3, therefore, was not applicable and the case was governed by a common law rule that Maker's payment to Peter did not discharge his obligation to Paul. Paul is wrong. Section 3–805 states that even though the writing is not an instrument, Article 3 controls on these facts. Section 3–116(a) states that Peter can discharge the instrument without Paul's consent. See 3–603 and Carpenter v. Payette Valley Co-op., Inc., 99 Idaho 143, 578 P.2d 1074 (1978). If the note were

REQUISITES OF NEGOTIABILITY Ch. 4

made payable to Peter and Paul, Section 3–116(b) would require a different result.

§ 4. Terms and Omissions Not Affecting Negotiability: 3–112

There are many terms that normally are not included that might be included without affecting negotiability. For example, negotiability is not destroyed by providing that by indorsing or cashing a check the payee acknowledges full satisfaction of some obligation owed to him. (For other examples, see Chapter 2 dealing with forms and uses of commercial paper.)

Also, many matters often mentioned in commercial paper may be omitted without destroying negotiability. For example, it is not necessary to mention the place where the instrument is issued or payable. Nor is it necessary to mention the consideration paid for the instrument even though its lack is a good defense against one who does not have the rights of a holder in due course. In fact it is not necessary to include any provision that is not expressly required by Section 3–104(1).

The negotiability is not prevented by including a term that purports to waive the benefit of a law intended to protect the obligor. 3–112(1)(e). Even when such waiver clauses are unenforceable, they do not destroy negotiability.

Nor is an instrument which otherwise meets the requisites of negotiability prevented from being negotiable by the fact that it is declared to be void because it contains a provision that is against public policy or because it arises from a transaction that is against public policy. This is so even though such a defense is good against a holder in due course. The reader might well ask what difference it makes that an instrument is classified as negotiable if it cannot be enforced even by a holder in due course. The answer is that the instrument may be acquired later by parties who are not involved in the illegality; and a determination of the rights and duties of these later parties is likely to depend to a large degree on the fact that the instrument is negotiable.

Suppose that the instrument contains a clause empowering the holder to confess judgment against the maker. As previously suggested, the fact that the clause may be against public policy does not preclude negotiability. Section 3–112(1)(d) provides that a term authorizing a confession of judgment on the instrument "if it is not paid when due" does not affect negotiability. The Code, however, does not change the earlier rule that authority to confess judgment "at any time after date" permits entry of a judgment prior to the maturity of the instrument and impairs negotiability. See, e. g., Von Frank v. Hershey Nat'l Bank, 269 Md. 138, 306 A.2d 207 (1973).

As a matter of interpretation, Section 3–112 should be regarded as a limited class of exceptions to the general statement that the writing contain "no other promise, order, obligation or power given by the maker or drawer * * * ". 3–104(1)(b). It should, therefore, be strictly construed.

Example 10

Maker issues to Payee an instrument to cover losses in a game of chance. Under state law, the game was illegal and Maker's obligation was a "nullity." If Payee negotiates the note to A who negotiates it to B who negotiates it to a holder in due course, none of these parties can recover from Maker on the instrument. Maker's personal defense will prevail. 3–305(2)(b). But only Maker can assert the defense of illegality, and the rights and duties of Payee, A, B and HDC among themselves are determined on the assumption that the note is negotiable.

Example 11

Maker signed a note dated July 1 promising to pay Payee "or order" the sum of $50,000 "60 days from date." The note authorized Payee or any Holder to accelerate the obligation "at any time he deems himself insecure" and to confess judgment if the note is "not paid when due." Payee negotiated the note to a HDC. Thirty days later, HDC, feeling insecure, accelerated the obligation and confessed judgment against Maker. Maker, who had defenses against Payee, sought to reopen the proceedings and vacate the judgment on the ground that the confession of judgment clause was illegal. If the clause is enforceable, the note is negotiable. HDC must confess judgment after acceleration not before maturity. See 3–109(1)(c) and 3–112(1)(d). Maker, however, might establish that the acceleration was in bad faith, 1–208, and that the confession of judgment was premature. If the clause is illegal, Maker can vacate the proceedings and insist upon a full hearing before judgment. The illegality, however, does not nullify the obligation, see 3–305(2)(b), and Maker would be unable to assert the defenses against HDC.

Ch. 4   *REQUISITES OF NEGOTIABILITY*

§ 5. Incomplete Instruments

What is the effect on negotiability of issuing an incomplete note or draft? Suppose that Drawer owes Payee's bakery $200 for goods delivered. Drawer draws a check in that amount but leaves blank the space provided for Payee's name. Drawer mails the check to Payee with the instructions to "fill in the blank with your trade name." The draft at this stage is incomplete and cannot be an instrument. 3–104(1)(d) and 3–110(1). Further, it cannot be enforced until completed. 3–115(1). When Payee completes the draft in accordance with the authority given, however, it is "effective as completed." 3–115(1). The check then becomes an instrument and is enforceable against Maker by Payee or any holder.

The problem becomes more complicated if Payee fills in the blank with the name of her spouse. The completion is unauthorized and a material alteration of the check. 3–407(1)(b). We will consider this general problem in more detail in Chapter 7, Section 2(G). For now please note that except as to a subsequent Holder in due course, a fraudulent material alteration by Payee will discharge Maker on the check. 3–407(2)(a).

§ 6. Omitting Date, Antedating, and Postdating

The negotiability of an instrument normally is not affected by the fact that it is undated. 3–114(1). If a person, inadvertently or otherwise, omits a date, there usually is no good reason to deny the instrument negotiability. If, for instance, it is payable "on July 1, 1982," a date for the instrument is immaterial. The

REQUISITES OF NEGOTIABILITY   Ch. 4

same is true if the instrument is payable "on demand." If a date is essential to fix maturity, however, as in an instrument payable "thirty days after date," the absence of a date prevents the instrument from being payable at a definite time and so bars negotiability until a date is inserted. 3–115; Comment 2, 3–114.

Negotiability is not affected by the fact that the instrument is antedated or postdated. 3–114(1). The time when postdated and antedated instruments are payable is determined by the stated date if the instrument is payable either on demand or at a fixed period after date. 3–114(2). For example, a demand note issued on May 1 but dated May 5 would not be due until May 5; and a note dated April 20 and stated to be payable one week after date but not issued until May 1 would be due when issued. The usual reason for issuing a postdated check is to obtain an extension of credit. By issuing such a check, a drawer promises in effect to have funds on deposit when the check falls due. It has been held that, unless there is an intent to defraud when the check is issued, issuing a postdated check knowing that there are not then sufficient funds on deposit to cover it is not a violation of a worthless check statute even though the check is dishonored for lack of funds on the date stated. Commonwealth v. Kelinson, 199 Pa.Super. 135, 184 A.2d 374 (1962).

Example 12

Drawer, father of an illegitimate child, drew a check for $20,000 in 1969 payable to the child's order. The check was dated July 1, 1985 but provided on the face that should Drawer die before date "this check shall be

Ch. 4   *REQUISITES OF NEGOTIABILITY*

payable immediately." Drawer issued the check to Payee's mother, the legal guardian. Drawer died on July 4, 1980. Despite the presence of sufficient funds, Bank refused payment upon due presentment. Similarly, Drawer's executor refused payment on the ground that the obligation was not due. Executor is wrong. A post-dated check is negotiable. 3–114(1). Although a post-dated demand instrument is normally payable at the stated date, 3–114(2), the acceleration clause made it payable upon Drawer's death. As such, the instrument was due and, in the absence of proof by Executor that there was no consideration, Payee, the holder of an instrument, is entitled to recover. See 3–307(2) and Smith v. Gentilotti, 371 Mass. 839, 359 N. E.2d 953 (1977).

§ 7. "Instrument," "Note," "Draft," and "Check" Presumed to Refer to Instrument That Is Negotiable

Although the study of commercial paper is concerned with writings that are not negotiable as well as with those that are, its primary concern is negotiable instruments. Therefore, in the discussion that follows it should be assumed that "instrument" refers to a negotiable instrument, rather than to one that is not, unless the context indicates otherwise. See Section 3–102(1)(e). This is consistent with the usage that prevails throughout Article 3 of the Code, and in the business community as well. For similar reasons, it should be assumed that "note," "draft," "check," or "certificate of deposit" refers to one that is negotiable rather than not negotiable. See Section 3–104(3).

CHAPTER 5

ISSUE: RELATIONSHIP BETWEEN THE IMMEDIATE PARTIES TO AN INSTRUMENT

The forms and requisites of a negotiable instrument have been discussed. Now we are concerned with its career in the business community. Its life span normally resembles that of a person. It is born, travels, dies, and is forgotten. For example, a check is born when it is issued, passes about by a process of transfer and negotiation, dies at the bank as the result of payment, and takes its place in the graveyard of cancelled checks. In this Chapter we are concerned primarily with only the first stage—the issue of commercial paper and the rights and duties of the *original parties* vis-a-vis each other. In later chapters, transfer and negotiation and the rights and duties of all of the parties—later holders and transferors as well as original parties—will be considered in their various combinations.

§ 1. Meaning of "Issue"

The signing of an instrument obviously is important: a writing cannot be negotiable unless it is signed by a maker or drawer. 3–104(1)(a). Also, a person cannot be liable on an instrument unless "his signature appears thereon." 3–401(1). Nonetheless, signing alone creates no liability. A maker or drawer, or anyone else, incurs no liability unless he delivers the instrument. The first delivery of an instrument by a maker or drawer to a "holder" is called *issue*. 3–102(1)(a). In this context, "holder" is defined as a "person," usually the payee, "who is in possession of * * * an instrument * * * issued * * * to him or his order or to bearer or in blank." 1–201(20). The delivery requirement is imposed by Article 3. Delivery is necessary to transfer title to the instrument, to constitute any person a holder and to impose liability upon signers of the instrument. In contrast, in the law of simple contracts, delivery may evidence an intention to be bound to a writing but is not required in all cases.

A maker or drawer who issues an instrument normally does so to create an obligation on the instrument to the person to whom it is delivered. The delivered instrument may create a new obligation or represent an existing one. Often the instrument is required by the terms of the underlying contract. For example, a contract for the sale of goods might provide that the buyer is to issue his 30-day note for the price. If the buyer signs the note but refuses to deliver it, he becomes liable for breach of the underlying contract, but he incurs no liability on the note. However, the fact that he has not delivered may not shield him from liability on the instrument if it is later stolen and delivered to someone who qualifies as a holder in due course. The fact that the note was not issued is no defense against a HDC. But the party in possession must be a holder, i. e., a party in possession of an instrument "issued" or "indorsed" to him. 1–201(20). Thus, if the stolen instrument is "order" paper, neither the thief nor any party taking from him can be a holder. More on this later.

§ 2. Meaning of "Delivery"

Delivery of an instrument is required for both issue, 3–102(1)(a), and negotiation. 3–202(1). Delivery means a "voluntary" transfer of possession. 1–201(14). Thus, a maker does not issue an instrument if he states to the payee "this note is yours" and retains physical possession. On the other hand, the voluntary transfer of physical possession from the maker to the payee is a delivery, whether or not the note is issued for consideration. In this setting, the maker's intention to transfer title is clear. But what of "constructive" delivery? Suppose that Maker makes a note payable to the order of Payee and delivers it to a third party, C. If Maker intended to pass title and C is Payee's agent or legal guardian, an issue has clearly occurred. But if either C's status or Maker's intention is unclear, questions of fact and law will arise. Frequently a rebuttable presumption of delivery will be invoked if the instrument is in the possession of someone other than Maker. Enough has been said to suggest that the question of when an instrument has been delivered with the requisite intention has not been fully answered by the Code and will be a continuing problem for the courts. See, e. g., Corporation Venezolana de Fomento v. Vintero Sales Corp., 452 F.Supp. 1108 (S.D.N.Y.1978).

§ 3. Effect of Delivery—Contract Between Immediate Parties

Between the immediate parties, a number of important consequences flow from the issue (or first delivery) of an instrument.

First, the holder has rights "on the instrument." He may enforce payment, discharge it or negotiate it. 3–301.

Second, when the instrument is taken for an underlying obligation, that obligation "is suspended pro tanto until the instrument is due or if it is payable on demand until its presentment." 3–802(1)(b). Thus, the instrument is the exclusive embodiment of the underlying obligation until the instrument is dishonored.

Third, the holder has certain procedural advantages in establishing signatures and avoiding defenses. 3–307.

Fourth, Article 3 governs the rights and duties of the parties. Thus, in enforcing the instrument against the maker or drawer, the holder must, among other things, (1) observe the statute of limitations, 3–122, (2) follow the appropriate notice and presentment requirements in Article 3, Part 5, and (3) observe the requirements for negotiation, 3–201.

Finally, the first holder is subject to a variety of claims and defenses established by the maker. See 3–307(2), 3–306 and 3–408. Article 3, therefore, extensively regulates the contractual relationship between the immediate parties.

§ 4. Interpretation and Scope of the Agreement

A. INTERPRETATION

In determining the rights and duties of the immediate parties after issue, one must consider the instrument and any other documents or writings that were executed at the time the instrument was issued. This accords with the principle that writings executed as part of the same transaction must be read together as a single agreement—at least as between the immediate parties. This is the substance of Section 3–119(1), which provides:

> As between the obligor and his immediate obligee or any transferee the terms of an instrument may be modified or affected by any other written agreement executed as a part of the same transaction, except that a holder in due course is not affected by any limitation of his rights arising out of the separate written agreement if he had no notice of the limitation when he took the instrument.

So long as they are part of the same transaction, the two writings may be read together even though they do not expressly refer to each other. Remember, however, that a separate writing "does not affect the negotiability of an instrument," 3–119(2), and that a holder in due course will probably take free of any limitations on the note contained in a separate writing. 3–119(1). See Northwestern Bank v. Neal, 271 S.C. 544, 248 S.E.2d 585 (1978).

As between the immediate parties, the existence of other writings executed as part of the transaction that produced the instrument have posed few problems.

Disputes over the meaning of terms which are part of the total agreement should be resolved under the rules of construction in Section 3-118 and the basic principles of contract interpretation. See, e.g., Restatement, Second, Contracts §§ 201-204.

Example 13

Payee, in a written contract dated July 1, sold Maker goods for a price, stated in figures, of $1,500. The writing clearly provided that the price was not payable if the goods were not delivered by August 1. Maker issued to Payee a negotiable note, dated July 1, promising unconditionally to pay $1,400 on August 1. The amount payable was expressed in words rather than figures. The goods were not delivered by August 1 and Maker dishonored the note. Payee sued Maker on the instrument. As between the parties, Maker should be able to show that the promise in the note was conditioned in the separate writing executed simultaneously with the note. The condition was part of the total agreement of the parties. Further, the conflict between the $1,500 price stated in figures and the $1,400 price expressed in words should be resolved in favor of the latter. 3-118(c). If, however, Payee had negotiated the note to a holder in due course, HDC would not be affected by the condition in the separate writing unless he had "notice of the limitation" when he took the instrument. 3-119(1).

B. SCOPE: THE PAROL EVIDENCE RULE

Problems of a different kind are posed when one of the immediate parties tries to show that the rights and duties arising from the instrument are somehow altered by an oral statement made at the time of issue or by a written or oral agreement made before it was issued. This would be the case in Example 13 if the written contract was dated July 1 and the instrument was dated and issued on July 15. The legal effect of these statements and agreements depends upon the parol evidence rule and, to a lesser extent, provisions of Article 3 which limit the freedom of the parties to introduce such evidence. See, e. g., 3–118. Here we are concerned with the effect of the parol evidence rule on such statements. If applicable, the "rule" operates to exclude them and thus narrow the scope of the agreement.

The parol evidence rule applies to negotiable instruments in much the same way that it applies to other written contracts.

According to the parol evidence rule—actually a rule of substantive law rather than a rule of evidence—when two parties reduce their agreement to writing, being under no mistake with respect to what the writing provides, with the intention of adopting the writing as the final and complete expression of their contract, neither party may thereafter introduce parol, or outside, evidence—that is, evidence of an oral agreement made at the time the writing was signed, or evidence of a prior oral or written agreement—for the purpose of showing that the terms they actually agreed upon were different from those expressed in the writing. It is reasoned that all of the prior or simultaneous agreements or understandings that are entitled to legal effect are embodied or merged in the one integrated writing, which is sometimes called an "integration," or "merger." In short, the parol evidence rule simply prevents either party from going back in time and contradicting or adding to the agreement. See 2-202.

The parol evidence rules applies not only when the parties adopt a writing as the final and complete expression of their *total* agreement but also when they adopt a writing as the final and complete expression of *only a part* of their agreement. In the latter case, the writing is called a "partial" integration. Depending on the facts, a writing in the form of a negotiable instrument may constitute either a complete or a partial integration of the agreement between the parties. In either case, its terms usually are protected by the parol evidence rule. Accordingly, with only minor exceptions which will be explained, neither the issuing party nor the party to whom the instrument is issued is permitted to introduce any evidence, written or oral, of what occurred prior to signing or any oral evidence of what was agreed at the time of signing for the purpose of changing, adding to, or deleting, any terms of the instrument. For example, such evidence would not be admitted for the purpose of changing the promise or order, the time or place of payment, the amount to be paid, or the words of negotiability. Similarly, parol evidence is not admissible to show that a promise unconditional on the face of the instrument was in fact conditional. Bank of Suffolk v. Kite, 427 N.Y.S.2d 782, 49 N.Y.2d 827, 404 N.E.2d 1323 (1980).

1. LIMITATIONS ON OPERATION OF PAROL EVIDENCE RULE

Although the parol evidence rule plays an important part in protecting an instrument against proof that it is not what it seems to be, there are a number of exceptions to its operation. These apply even though both parties intended that the instrument be either a total or partial integration of the agreement.

One limitation allows a party to have the instrument reformed to express what the parties intended to have it express, if he can prove by clear and convincing evidence that, as the result of a mutual mistake in reducing the agreement to writing, it provided something different from what they thought it provided when it was signed. See Comment 1, 3–118. Likewise, the parol evidence rule does not prevent a party from showing that *after* the instrument was signed, the parties agreed to change it. Nor does it prevent a party from introducing evidence to *avoid*, or set aside, an instrument, rather than vary it, by proving fraud, mistake, duress, undue influence, infancy or ordinary incompetency. Also a party who signs an instrument may show that he later was discharged from liability by an agreement or in any of the other ways recognized by law. Finally, he may introduce evidence to show that the instrument never took effect so as to impose liability on him because it arose from an illegal transaction or for some other reason. It will be noted that except for the limitation allowing reformation, none of the limitations admits evidence intended to vary the terms of the instrument.

Ch. 5 *ISSUE*

The relationship of the parol evidence rule to ambiguous terms merits special consideration. Sometimes, even though there is no mistake in reducing an agreement to writing, the words selected and used by the parties are ambiguous. The application of the parol evidence rule does not preclude evidence offered to resolve such ambiguity because such evidence merely explains and does not vary the terms used by the parties. Mainly to increase certainty in commercial dealings, however, the Code provides rules of construction to deal with several common ambiguities and to exclude parol evidence intended to resolve them even though such evidence might be admitted under the normal operation of the parol evidence rule. Most of the rules of construction provided by the Code deal with special matters which are discussed elsewhere in this book. It should be sufficient here to mention only two of these rules which are of a general nature. The first is that "handwritten terms control typewritten and printed terms, and typewritten control printed." 3–118(b). The second is that "words control figures except that if the words are ambiguous figures control." 3–118(c). Ambiguities which are governed by the rules of construction are resolved without giving any consideration to parol evidence. In short, although the Code, like the parol evidence rule, would not preclude parol evidence to reform an instrument the terms of which are, as the result of a mistake, not what the parties intended, and would admit parol evidence to resolve most ambiguities in the words chosen by the parties, it does preclude the admission of parol evidence to resolve ambiguities with respect to matters and terms that are governed by rules of construction. Comment 1, 3–118.

2. CONDITIONAL DELIVERY AND THE PAROL EVIDENCE RULE

Conditions Precedent and Conditions Subsequent. In applying the parol evidence rule to agreements made *at or before* the delivery of a negotiable instrument, the courts make a vital distinction between conditions *precedent* and conditions *subsequent.* A condition precedent is intended to prevent an instrument from taking effect until the occurrence of the condition. A condition subsequent is intended to terminate the liability of a party on an instrument *after* it has taken effect. Courts have consistently held that the parol evidence rule does not exclude parol evidence that an instrument was delivered on a condition precedent, but a large majority of the courts have held that it does exclude evidence of delivery on a condition subsequent. It is reasoned that proof of a condition precedent is not offered to contradict a contract; rather it is offered to prove that no contract ever came into existence. In contrast it usually is reasoned that proof of a condition subsequent assumes the existence of a written contract and is offered to show that although the promise contained in a writing appears to be unconditional, actually it is qualified by the condition subsequent. Since this would contradict the writing, it is excluded by the parol evidence rule.

Of course, there are many cases in which it is clear from the parol evidence that the parties agreed that the occurrence or non-occurrence of some event is intended to be a condition to the liability of the signing party, but it is difficult to determine whether the occurrence is a condition precedent to liability or the non-occurrence is a condition subsequent.

Typical cases in which a person who signs and delivers a negotiable instrument is *permitted* to prove by parol evidence the existence of a condition *precedent* that will shield him from liability are those in which a party signs a note as maker and delivers it with the understanding that he is not to become liable on it until some other named person signs as co-maker, or where a maker signs and delivers a renewal note with the understanding that it is not to take effect until the original note is returned, or where a buyer signs and delivers his check as part payment of the price with the understanding that it will not bind him unless and until he is able to obtain a loan for the balance from a third person. In these cases the evidence of the condition is introduced to show that the instrument never took effect as a contract, and not to contradict the writing, and so it is admissible. See, e. g., Weirton S. & L. Co. v. Cortez, 157 W.Va. 691, 203 S.E.2d 468 (1974).

Typical cases in which a person who signs and delivers a negotiable instrument is barred from introducing parol evidence of a condition *subsequent* are those in which the maker of a note offers to prove that, at the time of delivery, it was orally agreed that the maker would be discharged from liability by rendering some performance other than the payment of money or that, at the time of delivery, it was orally agreed that the maker would be liable only until some other specified person signed the note at which time the maker would be discharged. In cases such as these, the instrument is assumed to be an integrated writing that represents the true contract of the parties and the parol evidence is inadmissible because it is offered to change that contract.

3. UNDERSTANDING SIGNER IS NOT TO BE BOUND

Occasionally, a person sued on a negotiable instrument tries to avoid liability altogether by proving that, when he signed, it was agreed that he would not be liable *under any circumstances.* Various reasons have been given for such understandings. For example, it has been claimed that the instrument was signed as a mere formality, or was intended to conceal the identity of the true party or that it played a role in bookkeeping procedures. Sometimes it has appeared that there was a plan to deceive a third party such as a bank examiner. Obviously, to accept such a defense too readily would go far to undermine the stability of commercial paper. However, parol evidence offered to prove an agreement that a signer is not to be liable on an instrument under any circumstances seems to be no more contradictory of the terms of the instrument than parol evidence of a condition precedent, which is clearly admissible. Nonetheless, in a large majority of the cases, parol evidence of such an understanding has been excluded on the ground that it would contradict the writing and so violate the parol evidence rule. One rationalization of this result is that such evidence would contradict the writing by reducing the stated amount to zero. Britton, Bills & Notes 126 (Second Edition 1961). In some cases courts have asserted public policy as an additional reason for excluding such evidence. In most cases where evidence of an oral agreement that a signer is not to be bound at all is admitted into evidence despite the objection that it violates the parol evidence rule, it appears that the evidence is properly admissible to prove fraud or some other defense.

Perhaps the reader should be reminded that in this Chapter we have been discussing the rights and duties of the *original parties*. What has been stated would not necessarily apply with respect to a later party who is a holder in due course. For example, a party who is liable on an instrument normally would not be entitled to have the instrument reformed on the ground that a mutual mistake had been made in reducing it to writing, nor could he effectively assert a claim to an instrument if this would work to the detriment of a later holder in due course. Similarly, conditional delivery and most other defenses that could be asserted against an original party could not be asserted against a later holder in due course. (For a discussion of defenses that might be asserted against a holder in due course, see Chapter 11.)

§ 5. Similar Principles Apply to Subsequent Delivery

The foregoing discussion relates primarily to the *issue* or *first delivery* of an instrument and to the liability of the maker or drawer to the first holder. However, with exceptions applicable to a later holder who is a holder in due course, the same general principles apply in determining the effect of later deliveries that occur in the process of transfer and negotiation, the subject to which we now turn.

Example 14

Maker, a grain dealer, had received for storage and sale some 10,000 bushels of corn, grown by Payee, a farmer. Events caused Payee to doubt Maker's financial position. The parties met at a bank and, after negotiations, Maker issued a negotiable note to Payee promising to pay upon demand $25,000, the calculated value of the grain. At that time it was orally agreed that the note was issued for the "sole" purpose of securing Payee's grain and that Maker would not be liable on the note if the grain were destroyed "without Maker's fault or negligence." A tornado destroyed the storage bins and 50% of Payee's grain and he demanded payment on the note. Maker refused to pay and, in the following law suit, sought to introduce the following evidence: (1) the oral agreement that the note was for the limited purpose of security; (2) the oral condition on liability; and (3) evidence tending to show that both parties made a mistake in calculating the value of the grain and, thus, the amount promised in the note. If the note was intended by the parties to be an integrated writing, most courts would admit evidence that delivery was for a special purpose or that the parties had made a mutual mistake of fact. See, e. g., Brames v. Crates, 399 N.E.2d 437 (Ind.App.1980). But evidence of an oral condition upon the unconditional promise contained in the note is inadmissible under the parol evidence rule. Such evidence is clearly inconsistent with the unconditional promise. See Bank of Suffolk County v. Kite, 427 N.Y.S.2d 782, 49 N.Y.2d 827, 404 N.E.2d 1323 (1980). This outcome is effective against the immediate parties to the instrument but not against a holder in due course.

CHAPTER 6

TRANSFER AND NEGOTIATION

An important characteristic of the law of commercial paper is the ease with which it permits transfer from person to person of an instrument after issue. But whether the transfer is by assignment or negotiation, see 3-202(1), the subject is technical and the problems, at times, quite difficult. Yet mastery is worthwhile because correct answer to many of the problems which arise depend upon an accurate assessment of the status and rights of the transferee.

§ 1. Negotiation and Assignment

After issue, an instrument may be transferred by negotiation or by assignment. (For a discussion of assignments, see Chapter 3, Section 1.) Whether a transfer is by negotiation or assignment, a transferee normally acquires whatever rights his transferor had. This is the so-called "shelter" principle of property law embodied in the Code. 3-201(1). If a transfer falls short of being a negotiation, the transferee receives no more than his transferor had and does not enjoy the special advantages accorded a holder. If the transfer is by negotiation, the transferee becomes a holder. 3-202(1), 1-201(20). As a holder, whether or not he is the owner, he has the legal power to transfer the instrument further by assignment or negotiation; he normally has the power to enforce it in his own name 3-301; he can discharge the liability of any party on

Ch. 6 *TRANSFER AND NEGOTIATION*

the instrument by receiving payment or in a number of other ways (see Chapter 12) and he enjoys several procedural advantages. See 3–307. In addition, one who is a holder and satisfies the remaining requirements for being a holder in due course—he pays value and takes the instrument in good faith and without notice that it is overdue or has been dishonored or is subject to any claim or defense (see Chapter 10)—takes the instrument free of all claims and most defenses that might have been asserted against his transferor or any prior party (see Chapter 11) and enjoys the other benefits of being a holder in due course. In taking an instrument free of claims and defenses that might have been asserted against prior parties, a holder in due course acquires more than his transferor had. This idea that a negotiation might confer upon a transferee greater rights than were held by the transferor is a key feature of negotiability. For the above reasons, as well as others, it is important to know what is required in order to make a transfer that is a negotiation.

§ 2. Requisites of Negotiation

The essential ingredients of negotiation depend upon whether the instrument is payable to *bearer* or to *order* at the time of transfer. If it is payable to bearer, *delivery alone* suffices for a complete negotiation. If it is payable to order, a complete negotiation requires, first, an indorsement by the appropriate party, and second, delivery. 3–202(1). As a requisite of negotiation, "delivery" has the same meaning and is governed by the same principles applicable to the concept of issue. Basically it means a voluntary transfer of possession. See Chapter 5, Section 2.

§ 3. Bearer and Order Paper Distinguished

An instrument is payable to *bearer* if (1) it is payable to *bearer on its face*, see 3–111, and carries no indorsement by an appropriate party or (2) it carries one or more indorsements and the last of these made by an appropriate party is *in blank*. An instrument is payable to order if (1) it is payable *to order on its face*, see 3–110, and it carries no indorsement by an appropriate party or (2) it carries one or more indorsements and the last of these by an appropriate party is *special*.

§ 4. Blank and Special Indorsements Distinguished

To recapitulate, if an instrument is issued payable to the "order of bearer," the payee is a holder of bearer paper. 3–111(a). As such, the holder may exercise her power to negotiate the instrument by delivery alone. 3–202(1). No signature or "indorsement" is required and the transferor has no liability on the instrument. Unless she signs the instrument as an indorser, the risk that the maker or drawer will not pay falls upon the transferee. See 3–401(1) and 3–414(1).

If the instrument as issued is order paper, the payee can negotiate it only by delivery "with any necessary indorsement." 3–202(1). What is an indorsement? In the Code scheme, there are two types of indorsements, *blank* and *special*. A blank indorsement does not name an indorsee. 3–204(2). Usually it consists only of the indorser's signature. A special indorsement specifies a person to whom or to whose order payment is to be made, 3–204(1), and contains the signature of the indorser. Assume that Arthur, the payee of order paper, transfers it to Ben, who transfers it to Carl. On the reverse side of the instrument the following indorsements appear:

(1) Pay Ben (Signed) Arthur

(2) (Signed) Ben

Arthur's indorsement is special because it names the transferee. The same effect would occur if the indorsement had provided "pay Ben or order" or "pay to the order of Ben." The instrument remains "order" paper. 3-204(1). Ben's indorsement is blank because it does not name the transferee. Note that the instrument is now "payable to bearer" and Carl may negotiate it by "delivery alone." 3–204(2).

§ 5. Appropriate Party to Indorse

An indorsement can affect the character of an instrument as bearer or order paper or the character of a transfer as a negotiation or as an assignment only if it is made by an *appropriate* party. An indorsement by an inappropriate party must be ignored when considering these matters. (Such an indorsement, however, could impose liability on the indorser. This is so regardless of whether he signs his own name or forges someone else's name. (See 3–401(1).) The appropriate party to indorse is always the *holder* of the instrument or someone authorized to indorse on his behalf. 3–202(2).

A. HOLDER

A *holder* is a person who has *possession* of an instrument that *runs* to him. More precisely, a holder is in possession of an instrument "issued, or indorsed to him or his order or to bearer or in blank." 2–201(20).

A *bearer* instrument *runs* to any person who has possession of it. This is so whether he acquired possession by original issue from the maker or drawer, by delivery alone from the holder of a bearer instrument, or by delivery plus blank indorsement from the holder of either a bearer or an order instrument. *Even without delivery*, a finder or a thief of a bearer instrument is always a holder. A bearer instrument retains its character as bearer paper and runs to anyone in possession of it until it is indorsed specially; then it becomes order paper and runs to the special indorsee.

At the outset, an instrument which originates as *order* paper *runs* to the payee. If it is issued directly to the payee he becomes the *holder*. But if it is issued to someone else, that person is called the *remitter*; he is not, however, the holder. At this point there is no holder. When the remitter delivers the instrument to the person named as payee, the latter becomes the holder. If the payee indorses the instrument specially, it continues as order paper and runs to the special indorsee who, on obtaining possession, becomes the holder, and as such, the appropriate party to indorse. If the payee or special indorsee indorses the instrument in blank, it becomes a bearer instrument which runs to anyone who happens to acquire possession of

Ch. 6   *TRANSFER AND NEGOTIATION*

it, just as if it had been issued originally as a bearer instrument. This was recognized by Lord Mansfield in 1781 in deciding Peacock v. Rhodes, 2 Doug. 633, 97 Eng.Rep. 871, wherein he said, "I see no difference between a note indorsed blank and one payable to bearer. They both go by delivery and possession proves property in both cases." See 3–204(2). But remember, a thief of order paper cannot be a holder—it was not indorsed to him. Furthermore, no one taking the instrument from a thief can be a holder. That person must take by an indorsement "by or on behalf of the holder." 3–202(2). Since the thief is neither a holder nor an agent of the holder, transferees from him take only his nonexistent rights.

§ 6. Requisites of Negotiation Illustrated

The foregoing principles are illustrated by the following cases each involving a *negotiable* instrument.

Example 15

A bearer instrument is issued to P. Without indorsing, P delivers it to A. Result: The transfer to A was by negotiation and A became a holder. A bearer instrument can be negotiated by delivery alone and anyone in possession of a bearer instrument is the holder.

Example 16

P has an instrument which names him as payee. Without indorsing, P delivers it to A. Result: The transfer to A was an assignment and not a negotiation, and A does not become a holder, because the instrument runs to P. Negotiation of an order instrument requires indorsement by an appropriate party (P in this

[*104*]

case), in addition to delivery. A, however, has a specifically enforceable right to have the unqualified indorsement of P. 3–201(3).

Example 17

A bearer instrument is issued to P. T steals it from P and delivers it to B. Result: The transfer to B was by negotiation and B became holder. T did not acquire the instrument by negotiation because it was not delivered to him. However, a bearer instrument runs to anyone who acquires possession of it, even a thief. Therefore, T was a holder. As holder, T had power to negotiate the instrument to make B holder. This is so even though T, the thief, never was owner of the instrument.

Example 18

P has an instrument which names him as payee. T steals the instrument from P, indorses it in blank, and then delivers it to B. Result: Neither T nor B were holders. An order instrument can be negotiated only after it has been indorsed by an *appropriate* party. See 3–202(2). Since P, the payee, had not indorsed, no later party could become a holder. Notice also that since T was not the appropriate party to indorse, his indorsement could not affect the character of the paper so that it remained order paper that ran to P even though T's indorsement was in blank.

Example 19

P has an instrument naming him as payee. He indorses it in blank. It is stolen from P by T who sells and delivers it to B. Result: The transfer to B was by negotiation and B became a holder. Although the instrument started out as order paper, it became bearer paper when it was indorsed in blank by P, the appropriate party to indorse. Thereafter anyone who acquired possession of the instrument, even a thief, became holder and had the power to negotiate it by delivery with or without indorsement. If B had indorsed the instrument specially to C, the instrument would again have become order paper because it is the nature of the last indorsement by an appropriate party that determines the character of an instrument as bearer or order paper.

§ 7. "Holder" Not Synonymous with "Owner"

Perhaps it should be emphasized that although the holder and the owner are usually the same person, they frequently are two different persons. For example, a holder may be a thief, finder, agent, or other bailee who happens to have possession of an instrument that runs to him; and the owner may not be a holder because he never acquired possession, or because he gave it up, or because, although he has possession, the instrument does not run to him—for example, X purchases a note from the payee and obtains possession, but the payee neglects to indorse the note.

§ 8. Risks of Bearer and Order Paper

Bearer paper entails obvious risks. As we have seen, since it runs to whoever acquires possession, even a finder or a thief can become a holder and, as such, has the power to negotiate it to a third party so as to deprive the true owner of his property in the instrument. Also, as holder, he normally has the power to discharge any person from liability on an instrument by obtaining payment or satisfaction. All this is so even though the finder or the thief never becomes the owner. Because bearer paper carries much of the circulation potential and risks of money, it should be treated with the same caution usually exercised when handling money.

Fortunately the risks of bearer paper can be largely avoided by the use of order paper. The drawer or maker can start by issuing the instrument as order paper, and usually does. Also, the Code permits a holder of bearer paper to avoid the risks simply by indorsing the instrument specially to himself, thus converting the paper into order paper which thereafter can be negotiated only if he indorses it. 3–204(1). This is so, regardless of whether the paper originates as bearer paper or as order paper. If the instrument is payable to bearer because the last indorsement is in blank, the holder can convert it into order paper merely by inserting "Pay" followed by his own name directly above the blank indorsement, thereby converting it into a special indorsement. 3–204(3).

Whether the foregoing principles lead to justice in a given case is likely to depend on one's point of view. But it can hardly be denied that in most cases they lead to reasonably predictable results and so they are of great importance in the business world. For example, they mean that a drawer can safely mail his check to the payee because he knows that if it is stolen and the payee's signature is forged and the bank overlooks this and pays the check, it will not be allowed to charge his account because the drawer ordered the bank to pay the holder and the bank paid someone else. They mean that if a note is stolen from the payee or special indorsee who has not indorsed, the maker remains liable to the victim of the theft regardless of whether he pays the thief. They mean that if a good faith purchaser acquires an instrument relying on a forged indorsement of the payee, the innocent purchaser, and not the person whose signature is forged, must bear the loss. Usually, once the facts are determined these principles are relatively easy to apply. But there are two cases which pose more difficult problems because justice and expediency combine to require a *special* application of these principles. One case involves the *fictitious payee* and the other involves the *impostor*.

§ 9. Fictitious Payee Problem

The *fictitious payee* is a favorite device for those who practice skulduggery, especially the employee who wishes to swindle his employer by padding bills or payrolls.

Suppose that X, a merchant, gives Y, his employee, general authority to issue checks drawn on B Bank to pay creditors and employees. Intending to cheat X, and enrich himself, Y draws a check for $3,000 on B Bank payable to the order of F, and signs X's name as drawer. F is neither a creditor nor an employee, and Y intends F to have no interest in the check. Y indorses the check in the name of F naming himself as indorsee. Y promptly cashes the check at B Bank and retains the proceeds. B Bank charges X's account for the $3,000. When X learns of Y's duplicity, he demands that the bank recredit his account $3,000. When B Bank refuses, X sues B Bank.

Both X and B Bank agree that X was liable on the check and that B Bank had a right to charge X's account if, but only if, Y was a *holder*. See 4–401(1). X contends that Y was not a holder because the check was payable to the order of F, who was therefore the only appropriate party to indorse, and that it was never indorsed by F, so Y could not become holder. Logically, there is much to be said for X's position.

However, as a matter of policy, the Code favors B Bank. It does so by providing that "an indorsement *by any person* in the name of a named payee is effective if * * * a person signing as or on behalf of a maker or drawer intends the payee to have no interest in the instrument." 3–405(1)(b). (Emphasis added.) By virtue of this provision the indorsement of Y, the defrauder, (or his confederate or anyone else), in the name of F, the named payee, had, in these special circumstances, the same effect as if F, the named payee, had indorsed. Consequently, Y was in possession of an instrument that ran to him and therefore was the holder. When a person draws a negotiable draft or check he orders the drawee to pay the holder. Therefore when the bank in good faith paid Y, it was obeying X's order and so was entitled to charge his account.

Just as the above provision protects one who *pays* relying on a signature in the name of a named payee in these fictitious payee cases, so also it protects one who *purchases* such an instrument in good faith. Assume that instead of cashing the check himself, Y had promptly indorsed the check in F's name either in blank or naming H as indorsee and had sold the check for value to H. Assume further that the bank, acting pursuant to X's instructions, had refused to pay H when the latter presented the check for payment, and that H sued X on the check. By virtue of Section 3–405(1)(b), H would have qualified as a holder and if he met the other requirements he would have been a holder in due course and, as such, entitled to recover from X despite Y's fraud.

In the above case, the dishonest employee was the "drawer" of the check in the sense that he signed it with the authority of the employer who was named as drawer. Suppose that, instead of actually *signing* the check, Y merely *prepared* it for X's signature and that X signed it thinking that F was a business creditor. Or suppose that instead of preparing the check, Y merely prepared a payroll or other voucher on which F's name falsely appeared as creditor. In these situations, the Code again favors the drawee bank if it pays, this time by providing that "an indorsement by any person in the name of a named payee is effective if * * * an agent or employee of the maker or drawer has *supplied him with the name of the payee* intending the latter to have no such interest." 3–405(1)(c). (Emphasis added.) And, as shown above, it also favors a holder in due course who takes the instrument through the indorsement of the defrauder or anyone else.

Fictitious payee fraud may take many forms. Sometimes there is an actual person or company bearing the name of the payee, sometimes not. It makes no difference. Sometimes the defrauder is a corporate officer, such as the treasurer, sometimes he is only a clerk. Again, it makes no difference, although the corporate officer is more likely to be signing the instrument, and the clerk is more likely to be furnishing the name of the payee. The important factor is that the defrauder intends the named payee to have no interest in the instrument.

Behind the Code provisions governing fictitious payee frauds is the feeling that the risk of the employee's fraud in these cases should fall on the employer drawer or maker rather than on the paying good faith drawee or purchaser because (1) the employer is in a better position to avoid the risk by exercising proper care in the selection and supervision of his employees; (2) he is better able to protect himself by fidelity insurance; and (3) the cost of such insurance is properly an expense of the employer. Comment 4, 3–405.

Suppose that a bank acts in good faith but is *negligent* in paying a check indorsed in the name of a fictitious payee, does its negligence deprive it of the benefit of the fictitious payee principle? In Prudential Ins. Co. v. Marine Bank, 371 F.Supp. 1002 (E.D.Wis.1974) the court favored the bank and held that in a case such as this so long as the payor acted in good faith, it is intended by the Code that the loss should fall on the employer and that the payor's negligence is not relevant.

Example 20

Popeye Motors, Inc. is an important customer of Bluto, Inc., a supplier of auto parts. Olive was Popeye's bookkeeper and had no authority to sign checks on behalf of Popeye. In a thirty day period, the following transactions took place.

(1) A vice-president of Popeye signed a check to the order of Bluto to pay an actual debt. Before issue, Olive stole and altered the check, forged the indorsement and obtained payment from Drawee Bank. Section 3–405 is not available to protect Drawee against Popeye because the person signing the check "as or on behalf" of the drawer intended the payee to have an interest in the check. 3–405(1)(b).

(2) Olive handed the vice-president a blank check and said: "Please make this out to Bluto. We owe them $5,000 for parts." In fact, no debt existed. The vice-president complied. Olive then forged Bluto's indorsement and obtained payment from Drawee. Olive's indorsement in the name of Bluto is "effective" because she supplied Popeye with the name of the payee intending Bluto to have no interest in the instrument. 3–405(1)(c). Even though there was an actual payee there was no actual debt. This confirms Olive's intentions when the blank check was supplied.

(3) Olive, without authority, used a facsimile signature machine to impose the vice-president's signature on a check to the order of Bluto Auto Parts, Inc. There was no debt owed to Bluto and Olive knew it. Olive then stamped the back of the check "Bluto" and obtained payment at Drawee Bank. Olive, signing "as or on behalf" of the drawer, intended Bluto to have no interest in the instrument. 3–405(1)(b). The question is whether the indorsement by Olive "in the name of a named payee is effective." To limit the scope of Section 3–405, some courts have required the indorsement to be in the exact name of the payee. At the very least, Drawee must insist upon an exact indorsement. See Seattle-First Nat'l Bank v. Pacific Nat'l Bank of Washington, 22 Wash.App. 463, 587 P.2d 617 (1978). Under this view, the indorsement is not effective and

Drawee Bank is obligated to recharge Popeye's account.

§ 10. Impostor Problem

The fictitious payee problem discussed above usually arises from an *inside* job. In contrast, the impostor problem is almost always an *outside* job. Typically, a confidence man, by impersonating a respectable citizen and making some promise which he has no intention of keeping, induces a drawer or maker to deliver his check or note made payable to the order of the respectable citizen. After indorsing in the name of the payee, he practices his lowly art again to induce a second trusting party to purchase or pay the instrument. Finally, he departs with his loot, and the law must determine which of the two innocent and defrauded parties must bear the cost of his leech-like existence.

Before the Code the outcome of these cases usually depended on what the court found to be the *dominant intent* of the defrauded party. If he dealt face to face with the impostor, his dominant intent was usually found to be to deliver the instrument to the impostor. Consequently, the impostor was treated as *holder*, and the party who gave the instrument to the impostor bore the loss rather than the person who purchased the instrument or paid the impostor. If the parties dealt by mail or telegram, it usually was reasoned that the defrauded party intended to deliver the instrument to the person the impostor pretended to be. Consequently, the impostor did not become the holder and so the loss was borne by the person who purchased the instrument or paid the impostor.

The *dominant intent* test was criticized as a fiction because in the eyes of the deceived drawer or maker, the payee named and the defrauder are the same person so that there is only one intention, or if there are two, they are so intertwined as to be inseparable. The framers of the Code therefore rejected the test of dominant intent and refused expressly to distinguish between imposture face to face and by correspondence by providing that "an indorsement *by any person* in the name of a named payee is effective if * * * an impostor *by use of the mails or otherwise* has induced the maker or drawer to issue the instrument to him or his confederate in the name of the payee." 3–405(1)(a). (Emphasis added.)

Under the Code, regardless of how the imposture is carried out, when the impostor or his confederate or anyone else indorses the instrument it is as if the payee of an ordinary order instrument indorsed it. If the indorsement is in blank, the instrument immediately becomes payable to bearer so that the impostor or anyone else in possession of it becomes the holder and the proper person to negotiate it or to receive payment. If the instrument is indorsed specially, by anyone, and delivered to the special indorsee, the latter becomes the holder, and if he meets the other requirements, he becomes a holder in due course who is entitled to enforce the instrument against the defrauded maker or drawer. If a drawee pays the holder, whether he be the impostor or anyone else, the drawee is entitled to charge the drawer's account because a payment to the holder is in accordance with the drawer's order. The net result is that the ultimate loss is normally borne by the defrauded maker or drawer rather than by the transferee from the defrauder or by the drawee who pays the impostor or a transferee.

Even under the Code, however, if the defrauder, instead of misrepresenting himself to be another, misrepresents himself to be the *agent* of another, and thereby induces a maker or drawer to issue him a negotiable instrument made *payable to his alleged principal*, there can be no effective negotiation of the instrument unless the alleged principal, himself, indorses the instrument. Under the Code, the underlying reason for this result is the feeling that the maker or the drawer who takes the precaution of making the instrument payable to the principal is entitled to the principal's indorsement. Comment 2, 3–405. See Thornton & Co., Inc. v. Gwinnett Bank & Trust Co., 151 Ga.App. 641, 260 S.E.2d 765 (1979).

At this point, the reader might well ask whether an instrument that is payable to a fictitious payee or to an impersonated payee is payable to order or to bearer. It is arguable that it is payable to bearer, to order, or to neither. Perhaps it is best to recognize that such an instrument is anomalous and that although it is governed by some well established principles in a law of commercial paper it is treated in a way that cannot be reconciled with some other equally well recognized principles. The Code starts with a desire to help the good faith purchaser or drawee who pays. It might have done this simply by declaring such paper to be payable to bearer so as not to require any indorsement. But if it had done this it would have favored even those who did not insist on even the appearance of a regular chain of indorsements and it did not wish to go this far. Comment 1, 3-405. So it required that at least someone appear to sign on behalf of the person named as payee before anyone could become a holder of it.

In summary, Section 3-405 is a major exception to the general rule that an unauthorized payee indorsement of order paper is ineffective as a negotiation. As between the actual or purported maker or drawer and those who take from or pay the wrongdoer in good faith, the indorsement is effective if the conditions set forth are satisfied. The policy reasons for this outcome are similar to those involved when bearer paper is stolen and delivered to innocent third parties for value: given the regular appearance of the instrument to prospective purchasers or payors and the need to develop a secure market for commercial paper, the maker or drawer is deemed to be in the best position either to prevent the wrongdoing or to insure against it in the usual course of business. Thus, liability is imposed without consent or fault under this approach to risk allocation.

§ 11. Incomplete Negotiation

Occasionally, an order instrument is delivered for value by the payee or an indorsee without his indorsement. When this occurs, the transferee's status may be summarized as follows: (1) The transferee is not a holder. (2) The transferee acquires only the rights of the transferor against prior parties. (3) Unless the transferor and transferee have agreed otherwise, the transferee has the right to the *unqualified* indorsement of the transferor and if necessary may obtain a decree of specific performance ordering the transferor, under pain of punishment for contempt of court, to indorse. Because this order is issued by a court of equity, the transferee is sometimes said to have equitable title prior to obtaining the indorsement. Actually, he is the true owner even though he is not yet the holder. (4) When the transferor completes the negotiation by indorsing, the transferee becomes the holder. The transferee's status as holder is determined as of the time of the indorsement. 3–201(3) and Comment 7.

Example 21

M purchased corporate securities from P and issued a note to P in the amount of $25,000 for the price. The sale was induced by P's fraudulent representation that the securities had been properly registered under state law. P delivered the note without an indorsement to H, a good faith purchaser for value who had no notice of the fraud. H, as a transferee, is entitled to P's unqualified indorsement. If the indorsement is given before H has notice of the fraud, H becomes a holder in due course and takes free of M's defense. See 3–201(3), 3–302(1) and 3–305.

Suppose that H, a HDC, delivered the note without an indorsement to J who gave value but knew of P's fraud. Whether J obtains H's indorsement or not, he is protected by the "shelter" principle—he is entitled to the rights of his transferor, a holder in due course. If, however, J were a HDC in his own right and transferred the note back to P, the "shelter" would develop a fatal leak. P as a party to the fraud "cannot improve his position by taking from a later holder in due course." 3–201(1). See Rozen v. North Carolina Nat'l Bank, 588 F.2d 83 (4th Cir. 1978).

§ 12. Depositary Bank's Power to Supply Missing Indorsement to Complete Negotiation

Occasionally, through oversight or otherwise, a payee deposits a check or other item in a bank for collection without indorsing it. If ordinary principles applied, the bank could not become a holder without getting the depositor's indorsement. For the bank to return the item for indorsement in this situation, however, delays collection with little advantage to the drawer and none to the depositor or bank. Still, some drawers are more concerned with having the payee's indorsement as convenient evidence of payment than they are with speeding up the collection process. Reflecting a balance of these conflicting interests, Section 4–205(1) of the Code provides:

> A depositary bank which has taken an item for collection may supply any indorsement of the customer which is necessary to title unless the item contains the words "payee's indorsement required" or the like. In the absence of such a requirement a statement placed on the item by the depositary bank to the effect that the item was deposited by a customer or credited to his account is effective as the customer's indorsement.

Under this Section a depositary bank that receives an item for collection without the depositor's indorsement may indicate on the item that it has been credited to the depositor's account and this normally has the same effect as if the depositor had indorsed. The bank becomes a holder and, if it otherwise qualifies, it becomes a holder in due course. Also, if a drawee bank pays such an item, it is entitled to charge the drawer's account. However, if a check or other item shows that a payee's or indorsee's indorsement is required, no later party can become a holder without such indorsement.

§ 13. Negotiation by Multiple Payees

As stated earlier (page 71), an instrument is not prevented from being negotiable by the fact that it names several payees either in the alternative, as "A, B, or C," or together as "A, B and C." 3–110(1)(d).

If the instrument is payable to two or more payees in the alternative, it is payable to any one of them and it may be negotiated, discharged or enforced by any one of them who has possession of it. 3–116(a). To negotiate such an instrument, the indorsement of only one of them is required.

If the instrument is payable to several payees together, it may be negotiated, discharged, or enforced only by all of them acting together. 3–116(b). To negotiate an instrument that names several payees together, it must be indorsed by all of them. 3–116(b). Thus, in Cincinnati Ins. Co. v. First Nat'l Bank of Akron, 63 Ohio St.2d 220, 407 N.E.2d 519 (1980), the court held that a drawee bank breached the contract of deposit with the drawer when it paid a check payable to joint payees without obtaining the proper indorsement of both. And if the drawee pays A without obtaining the indorsement of B, a joint payee, the drawee will be liable to B for conversion under Section 3–419(1). Trust Co. of Columbus v. Refrigeration Supplies, Inc., 241 Ga. 406, 246 S.E.2d 282 (1978).

The naming of several payees together is often a convenient device for a person who owes one obliga-

Ch. 6 *TRANSFER AND NEGOTIATION*

tion to several persons. If all such named payees indorse an instrument to negotiate it or if all sign a receipt for payment, the distribution of the proceeds need not concern the party issuing the instrument. But the case in which one of such payees, allegedly without authority, has purported to act on behalf of the others, in signing or negotiating the instrument or in obtaining payment, has been the source of much litigation involving not only the payees, but also the issuing party and others as well.

§ 14. Effect of Attempt to Negotiate Less Than Balance Due

If an instrument has been paid in part it may be negotiated as to the balance due. To be effective as a negotiation, however, an indorsement must transfer the entire instrument or the entire unpaid balance. If the indorsement purports to transfer less, it operates only as a *partial assignment*, 3–202(3), and the partial assignee does not become a holder or acquire his advantages. Whether the partial assignee can sue to enforce his rights, and if so, the conditions under which he may do so, are left by the Code to local law. Comment 4, 3–202.

§ 15. Place of Indorsement on Instrument

Normally an indorsement appears on the reverse side of an instrument, but it may appear elsewhere. To cover cases where the capacity in which a person signed is uncertain the Code provides that unless the instrument clearly indicates that a signature is made in some other capacity, it is an indorsement. 3–402.

To provide for the unusual case in which there are so many indorsements that there is no room on the back for more, the Code provides that an indorsement may be written on the instrument "or on a paper so firmly affixed thereto as to become a part thereof." 3–202(2). A paper so affixed is called an *allonge*. See Shepherd Mall St. Bank v. Johnson, 603 P.2d 1115 (Okl.1979).

§ 16. Negotiation When Name Is Misspelled or Misstated

Occasionally an instrument is made payable to a payee or indorsee under a misspelled name or name other than his own. In this case, an indorsement in the name that appears on the instrument is legally sufficient for an effective negotiation, but this alone is commercially unsatisfactory because of the difficulty a later holder might have in proving the identity of the indorser. An indorsement in his true name alone also is legally sufficient for a negotiation, but this is unsatisfactory because a later transferee may be uncertain about the state of title. To avoid this difficulty, the Code provides that any person paying or giving value for the instrument may require the indorser to sign both names. 3–203.

§ 17. Effect of Negotiation That May Be Rescinded

If a transfer satisfies the described requirements, it is an effective negotiation even if it may be rescinded on the grounds of incapacity, fraud, duress, or illegality, or because it was made in breach of trust, 3–207(1), and even if it is made by a thief, a finder, or someone else acting without authority. In fact, although the transaction in which it occurs is held to be entirely void because of illegality, an adjudication of incompetency, or on any other ground, a transfer meeting the requirements is an effective negotiation. Comment 1, 3–207. Consequently, the transferee of an instrument in any of these transactions is a holder with full power to negotiate it further as long as he retains possession; and the transferor loses all of his rights in the instrument until he recovers it. Comment 2, 3–207. If the instrument falls into the hands of a holder in due course, even the right to reclaim it is lost. 3–305(1). Some of the defects and infirmities above, however, may be asserted by the maker against a holder in due course. 3–305(2). (For a discussion of claims and defenses, see Chapter 11.)

§ 18. Unqualified and Qualified Indorsements Affect Liability

As explained earlier, indorsements are classified as either blank or special to ascertain whether an instrument is bearer or order and thus determine if any indorsement is necessary for an effective negotiation of the instrument. Whether or not an indorsement is blank or special, it may be classified a second way—*unqualified* or *qualified*—to indicate the nature of the indorser's liability. An unqualified indorser's liability normally is of two types: (1) *secondary*, by which he is obliged to pay if the party expected to pay fails to do so and certain other conditions are met 3–414(1), and (2) *warranty*, by which he incurs liability if the instrument has been altered or if certain other special circumstances have arisen. 3–417. If he indorses in the usual way, by either blank or special indorsement, the indorsement is said to be unqualified, as the indorser incurs both types of liability. To disclaim his secondary liability, however, an indorser can include the words "without recourse," or words of similar meaning, in his indorsement; in which case, his indorsement, and consequently his liability, is said to be qualified. Comment 1, 3–414. (See Chapters 7, 8, 9.) A warranty, however, must be disclaimed by additional, clear language.

§ 19. Restrictive Indorsements

In addition to the general classification of "blank" or "special," an indorsement may be classified as unrestrictive or restrictive. An unrestrictive special indorsement, for example, limits the holder's method of future negotiation, see 3–202(1), but does not limit the transferor's liability as an indorser or how the holder should apply any funds received from payment. A restrictive indorsement, however, limits the rights and powers transferred in the ways specifically provided in Section 3–205.

Section 3–205 provides that "an indorsement is restrictive which either (a) is conditional; or (b) purports to prohibit further transfer of the instrument; or (c) includes the words "for collection", "for deposit", "pay any bank", or like terms signifying a purpose of deposit or collection; or (d) otherwise states that it is for the benefit or use of the indorser or another person." A very large proportion of the indorsements that are used in the business world outside of banking channels, are unrestrictive, a classification that presents no problems. The principal question raised by restrictive indorsements is how they affect indorsees, subsequent transferees, and payors.

A. INDORSEMENTS THAT PURPORT TO PROHIBIT FURTHER TRANSFER

Under the prior law there was uncertainty about when a restrictive indorsement prevented further negotiation. The Code removes all doubt by providing unequivocally that "*no* restrictive indorsement prevents further transfer or negotiation of the instrument." 3–206(1). (Emphasis added.) Even an indorsement which expressly prohibits further transfer cannot bar further transfer or negotiation under the Code. There are few practical reasons for a transferor to bar further transfer, and so such indorsements are rare. Under the Code, when they do occur they are given the same effect as unrestrictive indorsements. Thus, the indorsement, "Pay N. Dorsey only," is treated as if it were, "Pay N. Dorsey," or, "Pay to the order of N. Dorsey." This policy is consistent with the rule invalidating contract terms that prohibit the assignment of contract rights. See 9–318(4).

B. CONDITIONAL INDORSEMENT

Somewhat more prevalent than the indorsement purporting to restrict further transfer, but still not common, is the conditional indorsement. This imposes a condition on the right of the indorsee or any later holder to collect. Usually, the purpose of a conditional indorsement is to assure the indorser that some duty *owed to him* will be performed. For example, P, the payee of a note, may indorse it, "Pay X when he delivers 100 shares of Gold Stock to me. (Signed) P," and deliver it to X in return for X's promise to deliver the stock. A conditional indorsement should not be confused with a conditional promise by a maker or a conditional order by a drawer. A condition to a promise or order, if contained in the instrument, prevents the instrument from being negotiable. A conditional indorsement does not. And, as stated above, like other types of restrictive indorsements, it does not prevent further transfer or negotiation of the instrument. Under the Code, a conditional indorsement is treated in the same way as an indorsement for deposit or collection which is discussed immediately below. 3–206(1), (2), (3), 3–603(1)(b).

C. INDORSEMENT FOR DEPOSIT OR COLLECTION

By far the most important and common restrictive indorsements are those for deposit or collection. Typically, they are expressed, "for deposit," "for collection," or "pay any bank." Each of these indorsements should serve as a warning that a bank or other transferee might hold the instrument, at the outset at least, in trust or as agent for the restrictive indorser rather than as the owner. How such an indorsement otherwise affects persons who come into contact with such instruments depends largely upon the roles they play in the collection process.

Ch. 6 *TRANSFER AND NEGOTIATION*

If a seller in Boston receives a check or an ordinary draft which is payable to him and is drawn on his customer's bank in Denver, he might conceivably indorse the instrument "for collection," and forward it to his own agent in Denver with instructions to obtain payment and remit the proceeds. However, the seller in Boston usually will indorse the instrument "for deposit," then deposit it in his own bank in Boston, and let the bank attend to collection and to crediting his account with the proceeds. His bank will then indorse the instrument "for collection" or "pay any bank" and forward it through one or more banks to the Denver bank. There will be no actual transfer of funds corresponding to this specific transaction. Rather than transferring actual funds, the Denver bank will credit the presenting bank with the proceeds, this bank will credit the bank from which it received the instrument, and so on until ultimately the Boston bank will credit the seller's account if it has not already done so. In the above case, the Boston bank, the first bank to which the instrument was delivered for collection, is known as the *depositary* bank; the Denver bank, on whom the instrument is drawn, is known as the *payor* bank; and the other banks which participated in the collection process are called *intermediary* banks. 4–105.

1. DEPOSITARY BANKS AND TRANSFEREES OUTSIDE BANKING PROCESS

In a consideration of the effect of restrictive indorsements, depositary banks and transferees outside the normal banking process are treated alike. They are governed mainly by Section 3–206(3) which provides that "except for an intermediary bank, any transferee under an indorsement which is conditional or includes the words 'for collection,' 'for deposit,' 'pay any bank,' or like terms * * * *must pay or apply any value given by him for or on the security of the instrument consistently with the indorsement* and to the extent that he does so he becomes a holder for value. In addition, such transferee is a holder in due course if he otherwise complies with the requirements of Section 3–302 on what constitutes a holder in due course." (Emphasis added)

Suppose that P, the payee of a note, indorses it "for collection," and delivers it to A, his agent, with instructions to obtain payment from M, the maker. Typically, A will present the note to the maker at maturity, obtain payment, and remit the proceeds to P. In this case A becomes a holder, but not a holder for value because he gives no value. Since he is not a holder for value, A is not a holder in due course; thus, if M has any defense against P, he can assert it against A. If A had paid P the amount of the note when he received it, or by mutual agreement they had cancelled some debt owed by P to A, A would have given value, and if he met the other requirements, A would have been a holder in due course. Consequently, A would not have been subject to the defense which M might have had against P, unless A had learned of it prior to giving the value.

Suppose that A, instead of taking the note to M, transfers it to T; can T become a holder in due course? He can to the extent that he pays value for the instrument *consistently with P's restrictive indorsement*. If he actually pays A the amount of the note he becomes a holder for value despite the restrictive indorsement because such payment is contemplated by the indorsement itself. However, he does not pay value consistently with P's restrictive indorsement if he merely credits A *personally* with the amount of the note.

In Blaine, Gould & Short v. Bourne & Co., 11 R.I. 119 (1875), T drew a draft on C, naming himself as payee. After indorsing it in blank, T sold the draft to G. G indorsed it, "Pay A or order on account of G. (Signed) G," and delivered it to A. A then indorsed the draft, "Pay B. (Signed) A," and delivered it to B who obtained payment from C, the drawee, at maturity. B applied the proceeds of the draft on a debt owed to him by A. G sued B to recover the proceeds. G was awarded judgment. The restrictive indorsement gave B as a subsequent indorsee, notice of the fact that A was only an agent to collect and had no authority to apply the proceeds of the draft to pay his own debt. If the above case had been decided under the Code, the result would have been the same. By paying value by satisfying A's personal debt, B was not paying value "consistently with" G's restrictive indorsement.

These principles apply in substantially the same way in determining the rights and duties of the *depositary* bank which receives an instrument under a restrictive indorsement in the course of the normal banking process.

For example, if P, the payee, indorses a check, "for deposit," and delivers it to A, his agent, the depositary bank is charged with knowing that the proceeds must be credited to P's account. If the bank takes the check from A and pays A the amount of the check, or credits A's account or applies it toward satisfaction of A's indebtedness to the bank, the bank is liable to P for conversion because the value it gives is not consistent with P's restrictive indorsement. For the same reason, the bank could not qualify as a holder in due course and, if it sued the drawer, would be subject to any defense that the drawer might have. In contrast, if the bank credits the amount of the check to P's account and allows P to withdraw the proceeds, it could qualify as a holder in due course because in this case it applies the value it gives consistently with P's restrictive indorsement.

2. SPECIAL TREATMENT OF INTERMEDIARY AND PAYOR BANKS

Intermediary banks and payor banks must handle negotiable instruments, especially checks, in vast numbers and by assembly line methods. Their normal operations would be hampered greatly if they were required to stop and consider the effect of each individual indorsement in a chain of indorsements. Consequently, the Code provides that *intermediary and payor banks may ignore a restrictive indorsement of any person other than the bank's immediate transferor or the person presenting the instrument for payment.* (If a payor bank is also a depositary bank, it is classified as a depositary bank only, and is not given any special treatment.) 3–206(2), 4–205(2). This special immunity applies to all types of restrictive indorsements but is, of course, most often important in connection with indorsements for deposit or collection.

The need for this protection is demonstrated by the case of Soma v. Handrulis, 277 N.Y. 223, 14 N.E.2d 46 (1938). Plaintiff, payee of a check, indorsed it "for deposit," and was then induced by a defrauder to entrust it to him for "safe-keeping." The defrauder persuaded a third party to indorse the check in blank and deposit it in the latter's account with Globe Bank. Globe Bank indorsed the check for collection and delivered it to the Federal Reserve Bank. The latter obtained payment from West Chester Trust, the drawee bank, and paid Globe Bank, which permitted its depositor to withdraw the proceeds of the check. In holding that on those facts Federal Reserve Bank was liable to the payee, the New York Court of Appeals stated, "The indorsement was restrictive and prohibited further negotiation for any purpose except for collection and deposit in (payee's) account." If the *Soma* case had been decided under the Code, the result would have been different because the Federal Reserve Bank, as an *intermediary* bank, would not have been affected by any indorsement other than that of the Globe Bank, its immediate transferor, and so would not have been liable.

Neither the Globe Bank nor the West Chester Trust were made parties to the action. However, it is clear that under the Code the Globe Bank, as a *depositary* bank, would have been liable to Mrs. Soma; a depositary bank is affected by *all* prior indorsements and cannot become a holder in due course unless it pays in accordance with the terms of a prior restrictive indorsement. 3–206(3). It also is clear that, under the Code, West Chester Trust, the drawee, would not have been bound by the restrictive indorsement because, as a *payor bank*, it is bound by only the indorsement of the person presenting for payment. 3–206(2).

Several years after Soma v. Handrulis, in Leonardi v. Chase Nat'l Bank, 263 App.Div. 552, 33 N.Y.S.2d 706 (1942) another New York court applied Florida law to reach a result consistent with the Code. The payee deposited a check in a Florida bank after indorsing it for deposit. The depositary bank forwarded the check to defendant intermediary bank "for collection and credit." The defendant bank promptly obtained a final credit (which is equivalent to payment) for the amount of the check from the drawee. Then, learning that the depositary bank had become insolvent, defendant bank used the proceeds of the check to offset a debt owed it by the depositary bank by giving the depositary bank a "final credit" on its books. The payee sued to recover the amount of the check from the defendant intermediary bank, on the theory that defendant had converted the payee's funds to its own use. The court conceded that defendant intermediary bank had received the check as a subagent of the payee and continued as such until "final payment" was made to the depositary bank. However, it held for the defendant intermediary bank stating, "Credit is the exclusive means of payment in ordinary banking practice. 'The cash is not handled' and 'No money passes.' Here the

Ch. 6   *TRANSFER AND NEGOTIATION*

payment to the [depositary] bank was by means of a final solvent credit, as the result of which the agency terminated and the relationship of debtor and creditor ensued between defendant and the [depositary] bank." Under the Code, the result, but not the analysis, would have been the same. As an intermediary bank, the defendant would not have been affected by the plaintiff payee's restrictive indorsement. It would have been free to carry on its dealings with the depositary bank in their usual manner without regard to the depositor's restrictive indorsement.

Once an instrument is deposited for collection, it normally remains within banking channels until it is finally collected and credited. To deal with the unusual case, the Code provides that after an instrument has been indorsed, "pay any bank," or the like, "only a bank may acquire the rights of a holder (a) until the item has been returned to the customer initiating collection; or (b) until the item has been specially indorsed by a bank to a person who is not a bank." 4–201(2). Under this provision, if a depositary bank indorses an item, "pay any bank," as is typically the case, and it is later lost or stolen, the finder or thief normally would lack power to negotiate the instrument to anyone else. The above provision makes clear that outside of banking circles, an indorsement, "pay any bank," or the like, is a special indorsement so that the indorsement of a bank, as special indorsee, is necessary for further negotiation.

We will have more to say about the rights and duties of banks and customers in the check collection process in Chapter 13.

D. TRUST INDORSEMENTS

Sometimes the purpose of a restrictive indorsement is to benefit the indorser or someone else in some way other than by providing a condition to payment or by providing for collection or deposit. For example, P, the holder of a note, wishing to benefit himself, may indorse it, "Pay T in trust for P," or, wishing to benefit C, may indorse it, "Pay T for the benefit of C." Like other restrictive indorsements this does not prevent further negotiation of the instrument. But since the restrictive indorsee is a fiduciary, such an indorsement raises the question whether a person who takes the instrument from or through such an indorsee can qualify as a holder in due course. This question is answered by Section 3–206(4) which provides:

> The *first* taker under an indorsement for the benefit of the indorser or another person * * * must pay or apply any value given by him * * * consistently with the indorsement and to the extent that he does so he becomes a holder for value. In addition such taker is a holder in due course if he otherwise complies with the (usual requirements). A *later* holder for value is neither given notice nor otherwise affected by such restrictive indorsement unless he has knowledge that a fiduciary or other person has negotiated the instrument in any transaction for his own benefit or otherwise in breach of duty." (Emphasis added)

This provision is based on the assumption that the restrictive indorsee in a case of this kind is a fiduciary and, as such, owes a duty to use the instrument and its proceeds for the good of the beneficiary and not for his own advantage. But it also recognizes that trustees frequently and properly sell trust assets and that

Ch. 6   TRANSFER AND NEGOTIATION

a purchaser from the trustee may well act in good faith. It requires, however, that a person who deals with the fiduciary, himself, furnish value in a way that is not obviously going to benefit the trustee rather than the beneficiary. The provision also recognizes that after the instrument has been negotiated by the fiduciary, *later purchasers* should not be required to investigate that transaction. Therefore, the duty to act consistently with the restrictive indorsement is limited to the *first* taker. A later taker is not barred from being a holder in due course unless he *knows* when he takes the instrument that the fiduciary negotiated the instrument in breach of his duty.

Consider the following case.

Example 22

P conveyed land to M. In payment, M made and delivered two 5-year notes for $5,000 each. P indorsed each note, "Pay T in trust for C" and delivered it to T with the understanding that T would collect the notes at maturity and pay the proceeds to C, who was P's son. A month later, T indorsed and delivered both notes to X in payment of a $12,000 debt T owed X. X acted in good faith. A year later, X indorsed and delivered one of the notes to Y in payment of a debt of $6,000 owed by X to Y. At maturity X and Y demanded payment but M refused to pay either note on the ground the warranty of title contained in P's deed had been broken. This was a good defense against X. Even though X acted in good faith, and gave value, as the *first* taker he was not a holder in due course because he did not apply the value in a manner consistent with P's restrictive indorsement. As first taker, X was charged with notice that T negotiated the note in breach of his fiduciary duty. Y, as a *later* holder, was

[*134*]

under no duty to apply the value he gave in a manner consistent with P's restrictive indorsement nor was he charged with notice or otherwise affected by the restrictive indorsement unless he *knew* that T had negotiated the note to X in breach of his fiduciary duty. If X, instead of taking the notes in satisfaction of T's own debt, had paid T the value of the notes in cash, he would have paid value in a manner consistent with the restrictive indorsement and could have been a holder in due course.

E. EFFECT OF RESTRICTIVE INDORSEMENT ON DISCHARGE OF PAYOR

Normally when a party who is liable on a negotiable instrument pays the holder, he is discharged from his liability. There are several exceptions. (See pages 313–318). The only one which is of concern here is provided by Section 3–603(1)(b), which states in substance that unless a payor is either an intermediary bank or a payor bank, one who pays or satisfies the holder of an instrument that has been restrictively indorsed is not discharged from liability on an instrument, unless he pays in a manner consistent with the terms of the restrictive indorsement.

Example 23

M issues a note to P who indorses "for collection, P," and delivers it to A, P's agent. At maturity A demands payment and M pays A the amount of the note. M's liability is discharged. By paying A cash, M paid in a manner consistent with the terms of the restrictive indorsement. M's liability would not have been discharged if, instead of actually paying A, M had merely applied the amount of the note to satisfy a debt owed to M by A; for this would not have been consistent with P's restrictive indorsement.

§ 20. Words of Assignment, Waiver, Guarantee, Limitation or Disclaimer

In addition to the ways already considered, the legal consequences of an indorsement may be affected by words of assignment, waiver, guarantee, limitation, or disclaimer of liability. None of these words, however, prevent an indorsement from being effective for the purpose of negotiation, nor do they prevent further negotiation or prevent a transferee from becoming a holder or even a holder in due course. 3–202(4).

As we have seen, a negotiable instrument may be the physical embodiment of a whole series of dynamic transactions and the original issuance and subsequent transfer and negotiation may effect a whole chain of rights, duties, and liabilities of the various parties concerned. This should become increasingly clear as our discussion continues.

CHAPTER 7

LIABILITY OF PARTIES ON THE INSTRUMENT

§ 1. Various Bases of Liability

Ideally, the holder of an instrument should be able to see, by mere inspection of it, the chain of persons who are liable to him and the nature of the liability of each. The ideal however is rarely achieved. Some of the persons who come into contact with an instrument sign it, others do not. Some come into contact with an instrument as makers or drawers, others as acceptors, indorsers or sureties. Each of these parties assumes a different kind of liability. Moreover, the nature of the various obligations depends on whether or not the person seeking to collect has the rights of a holder in due course. Also, some indorsers incur liability on the instrument as well as on the basis of certain warranties, but other indorsers and those who transfer without indorsement become liable only on their warranties. In addition, parties who obtain acceptance or payment may incur liability on one or more warranties. Furthermore, a negotiable instrument usually is issued or transferred for the purpose of satisfying some underlying obligation the liability on which might or might not be suspended temporarily.

This involved network of rights and liabilities will be delineated in this and following chapters. The present chapter will be concerned primarily with the liability

Ch. 7 *LIABILITY OF PARTIES*

imposed only on those who *sign* an instrument—usually referred to as liability *on* the instrument. 3–401(1).

§ 2. General Principles Governing Liability on the Instrument

A person who signs an instrument is said to be liable *on* it, as maker, drawer, acceptor, or indorser. The liability incurred by each of these parties differs from the liability of each of the others. Before considering them separately, however, let us consider some general principles which apply to all.

A. PERSON'S SIGNATURE MUST APPEAR ON INSTRUMENT

The Code expressly provides that, "no person is liable on an instrument unless his signature appears thereon." 3–401(1). Consistent with this principle, an undisclosed principal, whose liability on a simple contract made on his behalf without disclosing his identity is well recognized, cannot be held liable on a negotiable instrument because his signature does not appear on it. Even when the principal's identity is disclosed orally, he does not incur liability on the instrument if his authorized agent signs only his own name. See 3–403(2)(a). Nor can a person become liable on an instrument merely by signing a separate writing expressing his willingness to be liable on the instrument.

Although the principle that no person can become liable on a negotiable instrument unless his signature appears on it is sound as a general proposition, it must be applied with caution and with an awareness that it sometimes requires the use of fictions, or at least a little stretching of ordinary language. As shown earlier, there are many ways in which a signature can be applied. And as will be seen in later discussion, the general proposition does not prevent imposing liability because an authorized agent signed on behalf of the person held liable, 3–403(1), or because a person who actually did not sign an instrument *is precluded* from denying an unauthorized signature as his own, 3–406, or because an unauthorized signature of another person operates as the signature of the unauthorized signer. 3–404(1).

B. HOLDER'S NAME NEED NOT APPEAR TO GIVE RIGHT TO SUE ON INSTRUMENT

One must not draw from the general proposition that no person is *liable* on an instrument unless his signature appears on it the inference that no one is *entitled to sue* on the instrument unless his signature appears on it. Not only is it not required that a person's *signature* appear on an instrument to entitle him to sue on it; it is not even required that his *name* appear on it. As stated earlier the proper person to sue on an instrument is usually the holder who might be the payee or a special indorsee, whose name appears on it but who has never signed the instrument, or a possessor of a bearer instrument whose name does not appear on it at all.

C. SIGNATURE

Previously considered in discussing the requisites of negotiability was the need to have the instrument signed by the maker or drawer. What was said there regarding what constitutes a signature applies not only to makers and drawers but to acceptors and indorsers as well. (See Chapter 4, Section 1B).

D. DETERMINING CAPACITY IN WHICH PARTY SIGNS

Normally it is clear that the person who signs on the face of an instrument in the lower right hand corner is either the maker or the drawer. It usually is equally certain that a signature on the reverse side is an indorsement and the drawee's signature vertically across the face of a draft is an acceptance. Occasionally a person signs in some unusual place, but his capacity becomes apparent when the instrument is considered as a unit. For example, a signature of John Smith in the lower left hand corner is obviously that of the maker of an instrument which starts, "I, John Smith, promise to pay." When for any reason the capacity of the person signing is ambiguous, the problem is governed by Section 3–402 which provides that "unless the instrument clearly indicates that a signature is made in some other capacity, it is an indorsement." Unless it is clear from the instrument that he signed in some other capacity, one who signs a negotiable instrument is an indorser. Parol evidence to the contrary is not admissible. Comment, 3–402. But see 3–403(2)(b). Sometimes a person whose signature appears on an instrument is not liable in any capacity. For example, a person normally would not be liable on the instrument if he placed "witness" after his name. Comment, 3–402. Nor would he be liable if he signed below "payment received" for these words show a purpose to give a receipt, not to indorse.

E. SIGNATURE BY AUTHORIZED REPRESENTATIVE

Under the general laws of agency, an agent normally cannot bind his principal without actual or apparent authority to do so. In our present discussion, we may assume that actual authority exists. But even assuming that an agent has authority to bind his principal as maker, drawer, acceptor, or indorser of a negotiable instrument, the legal consequences of his attempt to exercise his authority may vary considerably from case to case depending on how he proceeds.

For example, assume that P. Prince has given A. Axel authority to bind him as maker, drawer, acceptor, or indorser. If A. Axel exercises his authority in the usual way—by signing "P. Prince, by A. Axel, Agent," —there is no problem. As a matter of law, the principal is liable, the agent is not. Parol evidence to show a contrary understanding is not admissible. But Axel might undertake to exercise his authority by signing in any one of a number of other ways, each of which presents its own problem. Confining our discussion to four relatively simple examples, assume that the name of the principal does not already appear on the instrument and that A. Axel, intending to bind P. Prince and

Ch. 7 *LIABILITY OF PARTIES*

not himself, signs: (1) "P. Prince," (2) "A. Axel," (3) "A. Axel, Agent," or (4) "P. Prince, A. Axel."

If, intending to bind his principal and not himself, A. Axel signs "P. Prince," the legal consequences will be those intended—Prince will be bound and Axel will not. See 3–401(1) and 3–403(1). This type of signature, however, may present a problem to a third party or his transferee if he is later required to show the circumstances under which Prince's name was signed. The subsequent transferee, however, will have the benefit of a presumption that the signature was authorized. 3–307(1)(b).

If the agent signs his name "A. Axel," without explanation, he alone is bound on the instrument, 3–403(2)(a), even if he intends to bind only his principal, and the party with whom he deals understands his intention. Proof of an oral understanding to the contrary is barred even without the aid of the parol evidence rule. 3–403(2)(a). The principal is *not* liable on the instrument. This is in accordance with the general rule that no one is liable on an instrument unless his signature appears on it. 3–401(1). However, if the third party has a valid claim and cannot collect from the agent, he may recover from the principal on the underlying obligation for which the instrument was given.

If an authorized agent, intending to act only for his principal, signs his own name and immediately thereafter indicates his representative capacity, but does not show the name of his principal—for example, "A. Axel, Agent"—he may show in litigation between himself

[142]

and the person with whom he dealt, or someone having only the latter's rights, that, when he signed, it was understood by the parties that he acted only as agent and was not to be personally liable. 3–403(2)(b). If he is sued by a later holder in due course, however, the agent is personally liable and may not show that he acted only as agent. Regardless of who is seeking to recover, the principal is not liable on the instrument because his name does not appear on it. A holder who cannot collect from the agent, however, may recover from the principal on his underlying obligation.

If an authorized agent, intending to act only for his principal, shows his principal's name as well as his own but does not indicate that he signs in a representative capacity—for example, "P. Prince, A. Axel"—in litigation between himself and the person with whom he dealt, or someone having only the latter's rights, the agent again may show that when he signed both parties understood that he acted only as agent and was not to be liable personally. 3–403(2)(b). But if he is sued by a later holder in due course, he is personally liable and may not show that he acted only as agent. The principal, of course, is personally liable because his name was signed to the instrument with his authority.

In any of the above cases, if the agent is required to pay the third party, he is entitled to be indemnified by his principal because he acted with actual authority.

In general, the foregoing discussion of an authorized signing of a negotiable instrument would be equally applicable if, instead of representing P. Prince, an *individual*, A. Axel, the agent, had represented an *organization* which is broadly defined to include a corporation, partnership, government agency, estate, trust, or other legal or commercial entity other than an individual. 1–201(28), (30). However, the great variety of circumstances in which a legal representative might undertake to exercise his authority by signing a negotiable instrument would seem to dictate caution to anyone responsible for the legal aspects of such matters. Again, a safe signature model to bind the principal and avoid personal liability is: "Rexford Corp., by Phil Potts, Pres."

Caution is also dictated if one wishes to make certain that a representative who agrees to be liable individually is held to his agreement. As shown above, a representative who wishes to avoid individual liability can easily do so by showing the name of his principal and the representative capacity in which he signs and by inserting the word "by" before his signature. But one who wishes to bind the person or organization represented and also wishes to bind the representative individually cannot always rely on the fact that the representative does not use the word "by" or show his representative capacity. To be safe in these cases it would appear to be good practice to have the representative sign twice; first, showing his representative capacity; and second, before the word "individually." If this practice is followed, whatever disagreement there is in the matter is likely to occur before the signing rather than after a dispute arises and it becomes important to impose liability on the representative.

Example 24

P. Prince owed C. Cline $1,000, due in two weeks. Prince was off to Europe for a four week trip and authorized his friend, A. Axel, to make on his behalf a note for $1,000 to the order of Cline and payable in two weeks. Axel issued the note to Cline, signed it "A. Axel, Agent," and stated to Cline that he was acting on behalf of Prince. Cline then negotiated the note to D. Dempster for value and informed Dempster that Axel was acting on behalf of Prince. In two weeks, Dempster demanded payment from Axel. Prince was still in Europe. As between Dempster and Prince, Prince is not liable on the instrument. Even though Axel signed in a representative capacity, Prince's name does not appear on the instrument. 3–401(1), 3–403, Comment 2. As between Dempster and Axel, Axel is personally liable on the instrument * * * Dempster is not an "immediate" party and his knowledge of Axel's representative capacity is irrelevant. 3–403(2)(b). If, however, Dempster had sued and recovered from Cline as an indorser and Cline had sued Axel as a purported maker, Axel could establish his representative capacity and avoid personal liability. In that case, Cline could not recover against either Axel or Prince on the instrument and would have to sue Prince on the underlying obligation.

F. UNAUTHORIZED SIGNATURES

The Code makes no distinction between the signature made by an agent who is acting innocently but is without actual authority and the signature that is a forgery under the criminal laws. 1–201(43). Both kinds of signatures are classified as *unauthorized*. It is basic that an unauthorized signature is *inoperative as the signature of the person or other legal entity*

Ch. 7   LIABILITY OF PARTIES

whose name is signed unless the latter *ratifies* the unauthorized act or is *precluded* from denying it. This does not mean that the unauthorized signature has no legal effect. On the contrary, the unauthorized signature *operates as the signature of the unauthorized signer* in favor of any person who in good faith pays the instrument or takes it for value. 3–404(1). What has been said about an unauthorized signature is true not only when considering whether a person shall be liable on an instrument, the main question now being considered, but also in determining whether there has been an effective negotiation, which was discussed in Chapter 6.

The effect of ratifying an unauthorized signature is retroactive so as to (1) impose liability on the ratifying party as if he had signed in the first place and (2) to release the unauthorized signer from liability *on* the instrument. It does not necessarily relieve the latter of civil liability to the ratifying party, nor does it bar criminal action if the unauthorized signing amounts to a forgery. Comment 3, 3–404. Whether the unauthorized signature has been ratified by words or conduct is determined by the law of agency rather than Article 3, which is silent on the matter. See Carpenter v. Payette Valley Co-op., Inc., 99 Idaho 143, 578 P.2d 1074 (1978).

G. ALTERATION

Under the Code an alteration might consist of an addition, change or deletion; or it might consist of the completion of an incomplete instrument otherwise than as authorized. 3–407(1)(b).

Depending on several variables, an alteration might or might not result in discharging the liability incurred by anyone who had signed as maker, drawer, acceptor or indorser.

If the alteration is (1) by a holder, and (2) fraudulent, and (3) material in that it changes the contract of any party to the instrument, it results in a discharge of any person whose contract is changed by the alteration, unless that party assents or is precluded from asserting the defense. 3–407(2)(a).

If the alteration does not meet all three of the above conditions, it does not result in discharging the liability of anyone. The contract may be enforced according to its original terms and if the alteration consists of the completion of an incomplete instrument otherwise than as authorized, it may be enforced at least according to its terms *as authorized.* 3–407(2)(b).

Regardless of whether the above three conditions are met, if the instrument is acquired *after the alteration* by a *holder in due course,* he is entitled to enforce it according to its *original* terms if the alteration consists of an addition, change or deletion. And he can enforce it according to the terms of the instrument *as completed* if the alteration consists of the completion of an incomplete instrument otherwise than as authorized. 3–407(3). This is so even though the party who signed the instrument never issued it. One who signs an incomplete instrument assumes this risk.

Example 25

1. M buys goods from P and delivers P his note for the price payable to P's order. T steals the unindorsed note from P, raises the amount, and sells the note to A. Since T was not the holder, M was not discharged by the alteration. P is entitled to recover the note from A and enforce it against M according to its original tenor.

2. The holder of a draft that makes no reference to interest, mistakenly thinking that such an instrument bears interest at 12%, in good faith adds "with interest at 12%" following the amount. No one is discharged from liability. The alteration was not fraudulent. The instrument is enforceable, as originally provided, without interest.

3. M issues a note to P who indorses it "without recourse" and negotiates it to H. H fraudulently deletes "without recourse" from P's indorsement. P is discharged because his contract is changed. M is not.

4. P negotiates M's note for $1,000 to A who raises it to $10,000 and then negotiates it to H for $9,200. If H is a holder in due course, M and P are liable to H to the extent of $1,000, the original tenor of the note, and A is liable to H for $10,000. If H is not a holder in due course, he is not entitled to recover anything from M or P who were discharged by the holder's (A's) fraudulent alteration which changed their contracts, but A is liable to H for $10,000.

5. R signs a blank check, locks it in his safe, and instructs his clerk to complete and cash it when the payroll is determined. T steals the check, fills in the blanks to show himself as payee and a large sum as payable and cashes the check at B Bank. Assuming that B Bank is a holder in due course, it is entitled to the full amount of the check as completed. If not, it is entitled to nothing.

H. NEGLIGENCE CONTRIBUTING TO UNAUTHORIZED SIGNATURE OR ALTERATION

The foregoing principles that normally furnish a shield from liability or a basis for recovery on the ground of *unauthorized signature* or *alteration* are sometimes rendered inoperative by the fact that the negligence of the person who asserts the wrongdoing substantially contributed to it. The broad principle which has governed such cases for more than a century (see Young v. Grote, 4 Bing. 253, 130 Eng.Rep. 764 (1827)) is stated in a refined form in Section 3–406 which provides:

> Any person who by his negligence substantially contributes to a material alteration of the instrument or to the making of an unauthorized signature is precluded from asserting the alteration or lack of authority against a holder in due course or against a drawee or other payor who pays the instrument in good faith and in accordance with reasonable commercial standards of the drawee's or payor's business.

It should be emphasized that Section 3–406 applies only if there has been *negligence*. If a holder in due course acquires an instrument in which the amount payable has been raised from $500 to $5,000, his recovery will be limited to $500 if the defendant has not been negligent. But under this provision he will be entitled to the full $5,000 if the negligence of the defendant substantially contributed to the alteration. In contrast, it should be noted that if the alteration consists of completing an incomplete instrument (one that is signed while the space for some essential term is completely blank as where the space for the payee's name or for the amount is completely blank) contrary to authority, rather than by merely taking advantage of carelessness which makes it easier to *change* a term already inserted, a holder in due course is entitled to recover on the instrument as completed without regard to whether the party sued has been negligent.

It will be observed that Section 3–406 above protects not only a holder in due course but also a drawee or other payor who pays the instrument in good faith in accordance with reasonable commercial standards of the drawee's or payor's business.

For example, in Foutch v. Alexandria Bank & Trust Co., 177 Tenn. 348, 149 S.W.2d 76 (1941), the plaintiff habitually allowed payees to fill in the amount of checks before he signed them as drawer. On one occasion he signed a check in which the payee had inserted "18.00" well to the right of the dollar sign on one line and then, starting in the middle of the following line, had inserted "Eighteen and no/100" immediately in front of "dollars." After delivery the payee used the same pencil he had used to fill in the original amounts to insert a "4" immediately in front of the "18.00" and "Four Hundred" immediately in front of "Eighteen and no/100." He then cashed the altered check at the drawee bank which charged plaintiff's account. When the plaintiff sued to recover the $400 from the drawee bank, the court held that the plaintiff was precluded by his negligence from asserting the alteration. The court pointed out that the plaintiff's negligence had made it virtually impossible for the bank to detect the alteration.

Section 3–406 extends not only to cases where negligence substantially contributes to a *material alteration* but also to cases where negligence substantially contributes to an *unauthorized signature*. For example, a person who is in the habit of using a signature stamp or an automatic signing device might substantially contribute to a forgery by negligently allowing outsiders to have access to it, or a person might substantially contribute to a forgery by negligently mailing the instrument to the wrong person who has the same name as the intended payee but a different address. See, generally, White and Summers, Uniform Commercial Code 626–630 (2d ed. 1980).

§ 3. Liability of Particular Parties on the Instrument

A. IN GENERAL

A person who signs a negotiable instrument may become liable *on* it as a maker, drawer, acceptor, or indorser. Makers and acceptors are classified as *primary* parties. Drawers and indorsers are classified as *secondary* parties. 3–102(1)(d). In an effort to simplify the distinction, it is sometimes said that a person who is expected to pay is a primary party whereas one who backs him up is a secondary party. Although this statement contains some truth, it disregards the fact that the person expected to pay an ordinary draft or a check, by far the most common negotiable instrument, is a drawee, who is neither a primary nor a secondary party. In fact, as will be mentioned again later, a

Ch. 7   *LIABILITY OF PARTIES*

drawee, as such, is not a party to the instrument at all. Actually, a primary party—whether maker or acceptor—normally is both expected to pay and legally bound to pay without need for the holder to resort to anyone else first; whereas a secondary party normally is not expected to pay and is not legally bound to pay unless the party expected to pay fails to do so. But even these generalizations are not always sound, as will be seen.

B. MAKER'S LIABILITY

The maker engages that he will pay the instrument according to its "tenor at the time of his engagement" or, if it is incomplete upon issue, as completed if completed as authorized. 3–413(1). See 3–115(1). Since the maker is primarily liable on the note, unless he has been discharged or has some other defense, he must pay it at maturity. Normally he cannot look to anyone else for reimbursement when required to pay.

C. DRAWER'S LIABILITY

The drawer issues a draft, and like the maker of a note, he is liable only on the instrument and not on the basis of any warranties. Unlike a maker, however, the drawer does not promise to pay, at least not unconditionally; and he does not expect to pay. Instead he orders the drawee to pay. Consider the ordinary check. By issuing it, the drawer in effect says to the bank, "Pay the amount of this check to the holder when he presents it and charge my account." In effect, he says to the payee, "Take this check to the bank and ask for payment and if it does not pay, come back to me and I will." Basically, the drawer has "backup" liability. This is not stated in the draft he signs, which contains no promise but only an order expressed by the word "Pay" addressed to the drawee. In short, the drawer is only secondarily liable. In describing the drawer's secondary liability the Code states, "The drawer engages that upon dishonor of the draft and any necessary notice of dishonor or protest he will pay the amount of the draft to the holder or to any indorser who takes it up." 3–413(2). What constitutes dishonor and what is required in the way of notice or protest in order to fix the drawer's liability will be discussed later. (See Chapter 8.) Typically the holder or his agent presents the draft to the drawee for payment at maturity, the drawee fails to pay, and the holder thereafter mails notice of dishonor to the drawer. Although a drawer, like an indorser, may disclaim secondary liability by indorsing "without recourse," he rarely does so. Just as the maker's primary liability may be increased by any negligence on his part that contributes substantially to an alteration, so may the secondary liability of the drawer be increased. Also the risk of incurring greater liability than he authorized, imposed on the maker when he signs an incomplete instrument, is imposed on a drawer under similar circumstances.

D. DRAWEE'S LIABILITY

A drawee is not liable on a draft until he *accepts* it. 3–409(1). See 3–410(1). Therefore so long as he refrains from signing the draft and does not become an *acceptor* he incurs no liability *on it* to the holder as the result of refusing to pay or accept it. If the draft is accepted, the acceptor assumes the same liability as a maker on the instrument. 3–413(1).

A check or other draft does not *of itself* operate as an assignment of any funds in the hands of the drawee available for its payment. 3–409(1). This is so even though the drawee holds sufficient funds of the drawer to pay the draft and has made a contract with the drawer binding himself to the drawer to honor the draft when it is presented. Normally the payee is deemed to be only an incidental beneficiary of the contract between the drawer and drawee. Of course, a drawee who fails to honor a proper demand for payment might be held liable to the drawer if the failure violates a duty owed to the drawer on the basis of a contract of deposit or otherwise. See 4–401(1).

Example 26

Assume that E owes R $1,000 which is overdue.

1. R assigns his right to the $1000 to P. Since a claim for money is freely assignable, P is entitled to collect the $1000 from E upon notifying E of the assignment. E has no choice in the matter; he must pay P.

2. R draws a draft on E for $1000 payable on demand and delivers the draft to P. P presents the draft to E and demands payment. E refuses to pay P. P has no rights against E, even though E may be liable to R on any contract they had between them.

Although a check or draft does not of itself operate as an assignment of the funds in the hands of the drawee available for payment, the intent of the parties to effect an assignment may sometimes be inferred from other facts and, when the inference is clear, such instrument will be treated as a device for carrying out an assignment so as to render the drawee to whom it is addressed liable to the holder as assignee. Comment 1, 3–409.

Also, a drawee becomes liable for *conversion* if, having had a draft presented to him for acceptance or for payment, he not only refuses to pay or accept but also refuses to return the instrument despite a demand. 3–419(1)(a)(b). Although such a refusal does not constitute an acceptance, the result is much the same because he thereby becomes liable for the face amount of the draft. Depending on circumstances, and less frequently, a person named as drawee might become liable to a holder on other bases such as a letter of credit, fraud, promissory estoppel, breach of trust, or on the theory that the holder is a third party creditor beneficiary of the contract between the drawer and drawee.

E. ACCEPTOR'S LIABILITY

As stated earlier, the *drawee is not liable on a draft until he assents by accepting it* and becoming an acceptor. A bank's acceptance of a check is called a *certification*. (See pages 20–21) Under the N.I.L. it was possible to accept by signing the instrument or a separate paper such as a telegram or a letter. Because these *extrinsic* acceptances caused a degree of uncertainty inconsistent with the free transferability of commercial paper, the Code provides that the acceptance must appear *on the draft itself.* 3–410(1). *The acceptance may consist of a signature alone.* It is customary for an acceptor to sign vertically on the face of the draft, but a drawee's signature anywhere else normally is sufficient because usually the only reason he has for signing it is to accept. The acceptor may, and usually does, insert the word "accepted" above his signature. He may add the date, an important act if the draft is payable a stated period "after sight." In the latter case, if the acceptor fails to insert the date of his acceptance, the holder may supply the date in good faith and the inserted date will bind the acceptor. 3–410(3). Also the acceptor may indicate the place where the draft is payable.

As an exception to the general rule that a person does not become liable on an instrument until he both signs *and delivers* it, a person may become liable as an acceptor by signing and either delivering the instrument *or notifying* the holder of his acceptance. 3–410(1) and Comment 5.

The acceptor's liability on the instrument, like the maker's, is primary, and it also resembles the maker's liability in other ways. 3–413(1). If the instrument is complete when he accepts it, the acceptor engages that he will pay the draft according to the terms of the draft when he accepted it. Any later alteration will not increase his liability unless he either assents to the alteration or is precluded from asserting it. Similarly, if he signs an incomplete draft, he will be liable on the completed instrument if it is completed within the authority given by him, and he runs the risk of incurring greater liability than he authorized if the instrument is completed in an unauthorized manner and is later acquired by a holder in due course. Note that the acceptor does not necessarily agree to pay the instrument according to the terms as made out by the drawer but rather according to the terms at the time of his acceptance. If between issue and acceptance the instrument has been altered in any respect, the acceptor is liable on the altered instrument. He may have recourse against the presenter and one or more prior parties, however, as will be explained when liability based on warranties is discussed. (See Chapter 9).

Example 27

1. R draws a draft for $2,000. P, payee, obtains E's acceptance and negotiates the draft to T. T raises the amount to $5,000. The liability of R, E, and P is discharged. See 3–407(2)(a). However, if T negotiates the draft to HDC, E is liable primarily in the sum of $2,000, the tenor of the draft at the time of his acceptance. (R and P are secondarily liable for the same amount. T is liable for $5,000.)

2. R draws a draft on E for $2,000. P raises the amount to $5,000, then induces E to accept. P then negotiates the draft to HDC. E is liable to HDC for $5,000, the tenor of the draft at the time of his acceptance. (R is liable secondarily for $2,000. P is liable for $5,000.)

3. E accepts a draft in which the amount has not been filled in with the understanding that P, the payee, will fill in the proper amount as soon as he learns the price of the goods for which the draft is given. P learns that the amount is $2,000 but fraudulently inserts $5,000. If P sues E, he will collect nothing because E was discharged by the alteration. However, if P negotiates the draft as altered to HDC, HDC is enti-

tled to collect $5,000 from E. (The drawer's secondary liability to HDC is $5,000. P's liability also is $5,000.)

1. ACCEPTANCE VARYING DRAFT

A drawee unwilling to accept according to the terms of a draft as originally drawn may tender an acceptance which varies these terms. It may vary the amount payable, the time of payment, or any other term. Since the variance must be on the instrument itself, it is in a sense an alteration; but since the alteration is not made by a *holder*, it does not result in discharging anyone.

When a drawee proffers an acceptance varying the terms of the draft, two courses are open to the holder. He may refuse the acceptance and treat the draft as dishonored. If he does so, he gives up whatever rights he had against the drawee because the drawee then is entitled to have his varied acceptance cancelled. 3–412(1). On the other hand, the holder may assent and thereby acquire the obligation of the drawee to the extent of the varied acceptance. If the holder follows the latter course, each drawer and indorser who does not affirmatively assent to the variance is discharged. 3–412(3). The underlying reason for discharging a drawer or indorser who does not affirmatively assent is that he has agreed to be liable on a certain contract and should not be held liable on a different contract unless he agrees to be. The fact that the new contract may even be beneficial to the drawer or indorser is immaterial in this connection.

F. UNQUALIFIED INDORSER'S SECONDARY LIABILITY

A transferor is one who negotiates or otherwise transfers an instrument to someone else other than by original delivery. A transferor may or may not indorse the instrument. If he transfers an *order* instrument for *value* without any agreement to the contrary, however, the transferee is entitled to his unqualified indorsement. 3–201(3). If the instrument is payable to *bearer* when he receives it and there is no agreement to the contrary when he transfers it, his transferee cannot require him to indorse it. Of course, a prospective transferee may demand the transferor's indorsement before agreeing to purchase it. Whether or not he gets it is a matter of bargaining power, not of law.

An unqualified indorser acquires *secondary* liability on the instrument. 3–414(1). He also gives several implied warranties. Qualified indorsers (those who indorse "without recourse") and transferors who do not indorse give implied warranties, *but they incur no secondary liability*. Liability based on warranties will be considered in Chapter 9. Here we are concerned only with the secondary liability of the unqualified indorser, that is, his liability on the instrument.

The majority of transferors indorse without qualification and so become secondarily liable on the instrument. The liability of the unqualified indorser, like that of the drawer, does not become fixed unless certain conditions are satisfied. (See Chapter 8). He does not say, "I will pay." Rather he says, "I will pay

Ch. 7 LIABILITY OF PARTIES

if the instrument is dishonored and any necessary notice of dishonor is given and any necessary protest is made." Assuming that these conditions are satisfied, he agrees to be liable according to the tenor of the instrument at the time of his indorsement. 3–414(1). An indorser is not liable for any alteration in the instrument after it leaves his hands unless he assents to it or is precluded from asserting the alteration. On the other hand, if an instrument is altered before he transfers it, he is liable on it according to its altered terms even though he is totally unaware of the alteration when he transfers it.

Under the Code, an unqualified indorser's secondary liability runs to the holder of the instrument and any later indorser who takes up the instrument whether or not the later indorser was obligated to do so. 3–414(1).

Quite frequently there are several unqualified indorsers. Since the holder may immediately upon dishonor proceed against any one of them without proceeding against the others, it often becomes important to determine if the one who is required to pay the holder has any rights against the others. Such rights depend upon the order of priority among them. Except when they agree among themselves to the contrary, unqualified indorsers are liable to one another in the order in which they indorse, 3–414(2); that is, each indorser is liable to each later indorser. It is presumed that they indorsed in the order in which their signatures appear on the instrument, reading from the top. 3–414(2). But parol evidence is admissible to show that they indorsed in some other order or that they agreed to be liable in some order other than that in which they signed. Comment 4, 3–414. For example:

Example 28

H holds M's note which has been unqualifiedly indorsed by P (payee), A, and B in that order. At maturity the note is dishonored by M. H gives due notice to P, A, and B. H may proceed against any one of the indorsers on the basis of secondary liability. If H obtains payment from B, B may proceed against A, P, or M. If B obtains payment from A, A may proceed against P or M. If P pays he must look to M alone. However, if A and B contracted between themselves that A would be free of liability, as where A indorses for B's accommodation, B acquires no rights against A if B pays H; if A pays H, A is entitled to recover from P or M or from B.

G. QUALIFIED INDORSER'S LIABILITY

As mentioned earlier, a transferor who wishes to disclaim secondary liability, can do so by affixing a qualified indorsement. 3–414(1). He may signify his intention to sign as a qualified indorser by including the words "without recourse," which usually precede, but may follow, his signature. Although other words may be used to disclaim secondary liability, they rarely are. Regardless of whether the disclaimer of secondary liability is expressed by the words "without recourse" or by other words, it must appear on the instrument itself; it cannot be proven by parol evidence because the disclaimer varies the terms of the written contract of indorsement. Comment 1, 3–414. Although a qualified indorser incurs no secondary liability, he incurs liability on warranties as do unqualified indorsers and those who transfer an instrument without any indorsement. With one exception, which will be explained, the warranty liability of a qualified indorser is the same as that of an unqualified indorser and one who transfers without indorsement. See 3–417(3).

Example 29

P, a dealer, sold a used car to M, who was 16 years old. Under state law, M's infancy was a defense to a simple contract. M made and issued a note to P's order in the amount of the purchase price. P negotiated the note to H, a HDC. Thereafter, H negotiated the note to J, a HDC. H signed a special indorsement "without recourse" on the back of the instrument. M, upon presentment, dishonored the note and asserted the infancy defense. P is now insolvent. All concede that M's defense is valid against a holder in due course. 3–305(2)(a). As between H and J, H, by the qualified indorsement, has disclaimed his secondary liability on the instrument. 3–414(1). A transfer "without recourse," however, does not disclaim any warranties that H made to J when the instrument was transferred for value. It does, however, modify the usual warranty made by the transferor that "no defense of any party is good against him," 3–417(2)(d): the warranty is limited to a warranty that he has "no knowledge of such a defense." 3–417(3). Unless H knew that M was an infant, therefore, H has no warranty liability to J. Warranties under Section 3–417 can be disclaimed between the parties by an explicit agreement which appears on the instrument. An indorsement, "Pay to J without recourse and without warranties, express or implied." (S)H," would be effective.

H. LIABILITY FOR CONVERSION

For many years the law of torts has recognized that recovery might be had for conversion of intangible property that is evidenced by a document such as a negotiable instrument which gives a large measure of control over the property. Reference has been made to liability for conversion that might be imposed on a drawee who refuses to return on demand a draft presented for acceptance or payment. (See page 155) A maker, drawer, or acceptor, or indorser also may be held liable for refusing on demand to pay or *return* an instrument presented for payment. 3–419(1)(b). In addition, any party may be held liable for conversion if he pays an instrument on a forged indorsement. 3–419(1)(c).

Example 30

M makes and delivers a note to P, for value. T steals the note from P, forges P's indorsement and then sells the note to I, an innocent purchaser. At maturity, I obtains payment from M who takes up the note. P is, of course, entitled to recover for conversion from T, the thief, or from I, the innocent purchaser. P is also entitled to recover for conversion from M, who paid the note on P's forged indorsement. Even though M paid in good faith, M's payment is an exercise of dominion and control over the note inconsistent with P's rights as owner. Comment 3, 3–419.

§ 4. Special Kinds of Liability on the Instrument

The special nature of the liability of one who indorses "without recourse" has been mentioned. Special kinds of liability also may arise when one person signs for the accommodation of another or when he transfers an instrument using an indorsement which contains words of guarantee.

A. ACCOMMODATION AND ACCOMMODATED PARTIES

Normally a person who signs a negotiable instrument does so as part of a transaction in which he has a major interest. In contrast, an *accommodation* party signs an instrument to lend his name to some other party to it. 3–415(1). The person to whom the accommodation party lends his name is the *accommodated* party.

An *accommodation* party may sign as maker, acceptor, drawer or indorser. Usually, he signs as indorser or co-maker. However he signs, he becomes liable in that capacity except that he does not become liable to the party accommodated. 3–415(2), (5). If he signs as a maker or acceptor, he is a primary party and is liable to the holder immediately when the instrument becomes due; he has no right to require the holder to present the instrument to the accommodated party first. 3–415(2). If he signs as an indorser or drawer he is secondarily liable and is entitled to require presentment, dishonor and notice of dishonor before he pays. (See Chapter 8)

The accommodation party might or might not receive consideration for lending his name. The Code does not require that he receive consideration in order to be bound in the capacity in which he signs. As all parties expect, any consideration furnished by the creditor moves to the accommodated party. In the typical case, where the accommodation party signs before the instrument is delivered, for the purpose of inducing the creditor to extend credit to the accommodated party, the accommodation party's promise is supported by the same consideration that supports the accommodated party's promise. Often, however, the accommodation party signs an instrument after it has been delivered and after the creditor has already furnished the consideration to the accommodated party. For example, the creditor makes a loan to the accommodated party and takes the latter's note, and thereafter the accommodation party signs as an unqualified indorser. When this occurs, the fact that the creditor does not agree to extend the time of the loan or to furnish any other consideration does not shield the accommodation party from liability. All that is required to render the accommodation party liable in the capacity in which he signs is that the creditor shall have taken the instrument for *value* before it is due. 3–415(2). What is required in order to take an instrument for value is explained more fully when the requisites for being a holder in due course are discussed. (See Chapter 10, Section 5) It is sufficient here to know that in the above case the creditor gave value when he made the loan because *executed consideration* is always value. It is important to note, however, that an *unperformed promise*, perhaps the most common form of consideration, normally does not constitute value for this purpose. 3–303(a). Compare 1–201(44)(d).

The distinguishing characteristic of the accommodation party is that he is always a *surety*. Comment 1,

Ch. 7   *LIABILITY OF PARTIES*

3–415. The accommodated party is his *principal*. Therefore the accommodation and the accommodated parties and those who deal with them as such are governed by general principles of suretyship law. Basically an accommodation party differs from other sureties only in that his liability is on the instrument and he is surety for another party on the instrument. Comment 1, 3–415. Accordingly, an accommodation party is not liable to the party accommodated and, if he pays the instrument, he has a right to *reimbursement* from the party who is accommodated. 3–415(5). Also he usually has a right of *exoneration*, and *subrogation*, and, if there are cosigners, he usually has a right of *contribution*. *Exoneration* entitles him to require his principal, the accommodated party, to pay when the instrument falls due even though the accommodation party is a maker or acceptor and the accommodated party is an indorser. Once he has paid the creditor in full, *subrogation* entitles him to take the place of the creditor and to have the creditor's rights against the principal and any security held by the creditor. If there are two or more co-signers, *contribution* entitles him to require the remaining co-signers to pay their share if he pays more than his share. Frequently the accommodation party does no more than sign his name and his various rights are declared by the courts in accordance with well established principles and the probable understanding of the parties. Most other sureties enter written contracts that state many of their rights in some detail.

Example 31

Without receiving consideration from P, M makes and delivers a 90-day note payable to P to enable P to purchase a truck from C. P negotiates the note to C by unqualified indorsement and C delivers the truck to P. As security for the note, P delivers stock to C. After maturity, without first having demanded payment from P, C demands payment from M. M is liable to C on the note. Even though M signed as accommodation maker for P, M is liable in the capacity in which he signed and so has no right to insist on presentment to P first. It makes no difference that M received no consideration from P. To render M liable as maker, it is sufficient that C gave value, which he did by delivering the truck. Since M is a surety, if he makes any payment to C, he is entitled to reimbursement from P. If M pays C in full, M is entitled to be subrogated to C's rights against P as well as to C's rights with respect to the security. This latter right is especially valuable if P is insolvent. If C had first proceeded against P on P's secondary liability, P could not have complained because P had no right to expect M to pay the note; and if P had been required to pay C, he would not have been entitled to reimbursement from M.

B. WORDS OF GUARANTEE IN AN INDORSEMENT

It has been indicated that for purposes of negotiation an indorsement's effectiveness is not altered by the fact that the indorsement contains words of guarantee. (See Chapter 6, Section 2) However, such words may substantially affect the liability of the indorser.

If an indorser adds to his signature the words "*payment* guaranteed" or other words having the same meaning, he engages that if the instrument is not paid when it falls due, he will pay immediately just as if he were a co-maker. 3–416(1). Unlike the usual unqualified indorser, one who indorses in this way has no right to insist on either presentment for payment to the maker or drawee or notice of dishonor. 3–416(5). He is liable to a holder immediately upon maturity without any conditions precedent. If he instead adds the words "*collection* guaranteed," he again waives any requirements that otherwise would have existed with respect to presentment or notice of dishonor, but he adds some conditions to his liability that would not have existed. He does not say, "I will pay." He says, "I will pay if the instrument is not paid when due, provided (1) the holder reduces his claim to judgment and execution is returned unsatisfied, or (2) the maker or acceptor has become insolvent, or (3) it is otherwise apparent that it is useless for the holder to proceed against the maker or acceptor." 3–416(2). Words of guarantee which do not specify otherwise guarantee *payment*. 3–416(3).

Example 32

M applied to P to be a dealer in P's products. Dealers were required to sign a contract containing, among other things, a promise to pay for products supplied on a continuing basis, and a time note promising to pay for the initial inventory. P would not approve the application unless M provided a surety on the note and contract. F, M's father who had no interest in the dealership, was then persuaded to sign the note with

M and to guaranty payment under the contract. Shortly after commencing, M, due to mismanagement, ceased business operations and vacated the premises. When the note was not paid on time, P sued M and F jointly on the note. Shortly before trial, M declared bankruptcy and P amended its complaint to exclude M and leave F as the sole defendant. A verdict for P was entered on the note and F appealed.

The judgment should be affirmed. Although F signed the note as an accommodation party or surety, he is liable in the capacity, i. e., co-maker, in which he signed. Thus, F was a surety who undertook primary liability on the instrument without regard to P's prior demand against M. As a surety, however, F will be discharged from liability if P has unreasonably and without authorization materially altered the risk that F reasonably assumed in his commitment. Discharge in these cases is justified by the ancillary nature of the surety's contract. Bankruptcy of M, the principal debtor, however, is not a defense that F can assert to justify discharge. Bankruptcy is part of the credit risk assumed and the decision of P not to pursue its action against M cannot be held to have materially altered that risk. See 3–415(3). Thus, F is liable on the note as maker and must pursue his remedies, if any, in subrogation or contribution against M. See American Oil Co. v. Valenti, 179 Conn. 349, 426 A.2d 305 (1979).

§ 5. Accrual of Cause of Action

When does a cause of action accrue against the various parties who are liable on an instrument? The answer is found in Section 3–122, a section not incorporated in the earlier Uniform Negotiable Instruments Law. It turns on the nature of the party's liability and whether the writing is a demand or a time instrument. When is an accrued cause of action barred by the statute of limitations? The answer is not found in the Code. Other relevant state statutes must be consulted in each case. For example, Section 3–122(1)(b) provides that the cause of action against the maker of a dated demand note accrues "upon its date." The applicable non-code statute of limitations for contract claims may provide that an action must be brought within six years from the date of accrual or be barred.

A. AGAINST MAKER OR ACCEPTOR

The maker or acceptor engages that he will pay the instrument "according to its tenor at the time of his engagement." 3–413(1). If payment is to be upon demand, the cause of action accrues upon the date of the instrument or "if no date is stated, on the date of issue." 3–122(1)(b). Thus, if the demand note were issued on May 15 and dated June 1, the cause of action would accrue on June 1 without any demand for payment by the holder. Whether or not the instrument is payable on demand may be the subject of some dispute. For example, in Seattle-First Nat'l Bank v. Schriber, 282 Or. 625, 580 P.2d 1012 (1978), the note, stated to be "due" on June 13, 1970, was dated December 15, 1969 and was payable "on demand but not later than 180 days after date." The trial court held as a matter of law that the note was a demand instrument and that a suit filed more than six years after the date of the note but within six years of the stated "due date" was barred by the statute of limitations. In reversing and remanding for a new trial, the court held that the "not later than" language created an ambiguity as to what the parties intended. This ambiguity could not be resolved as a matter of law under either Section 3–108 or Section 3–118.

If the instrument is payable at a stated time, i. e., a "time" instrument, see 3–109, the cause of action accrues on the "day after maturity." 3–122(1)(a). See Mechanics Nat'l Bank of Worcester v. Killeen, 377 Mass. 100, 384 N.E.2d 1231 (1979) (no default until the end of the day after maturity). Suppose the instrument is payable in stated time installments. If the maker defaults in the first installment, does a cause of action accrue at the end of the next day or must the holder wait until full maturity? The common law rule was that even without a default the cause of action accrued on the day following the time for payment of each installment. Section 3–122(1)(a) does not speak squarely to installment notes, but some courts have held that the common law rule was not displaced. See 1–103. Rather, it supplements Section 3–122 and insures that the cause of action accrues as each installment comes due. See Avery v. Weitz, 44 Md.App. 152, 407 A.2d 769 (1979).

B. AGAINST DRAWER AND INDORSER

Because the liability of a drawer or indorser on the instrument is secondary rather than primary, 3–413(2) and 3–414(1), the cause of action against them does not accrue until "demand following dishonor of the instrument." 3–122(3). Notice of dishonor is a demand. A similar rule applies to certificates of deposit, whether a demand or a time certificate. The cause of action does not accrue until demand, but "demand on a time certificate may not be made until on or after the date of maturity." 3–122(2).

Frequently, indorsers assent to "waivers" of presentment, demand, protest and notice of dishonor in a clause printed on the back of the note. These waivers eliminate some of the conditions to secondary liability and, to that extent, operate to the indorser's disadvantage. A waiver of demand, however, is beneficial because if the cause of action accrues without a demand, the chance that the statute of limitations will ultimately bar the claim is increased.

C. AGAINST GUARANTOR

If a payee of an instrument signs the back and delivers it to a holder, he assumes the liability of an indorser. 3–414(1). See 3–402. If the payee signs and adds the words "payment guaranteed," however, he engages that upon the maker's default he will pay "without resort by the holder to any other party." 3–416(1). When does a cause of action accrue against a guarantor, when it accrues against the maker or when the maker defaults in payment? If the maker is the party who signs the back as a guarantor, the answer is clear. Section 3–416(4) provides that no "words of guaranty added to the signature of a sole maker or acceptor affect his liability on the instrument." Presumably, these words will not affect when the cause of action accrues. The case of the third party guarantor, however, is different. Whether a demand or time instrument is involved, the guarantor is not exposed to liability until the maker dishonors. The maker is exposed immediately in a demand instrument and the day after the stated time in a time instrument. Under this analysis, the guarantor should be treated like an indorser rather than a maker or acceptor for purposes of Section 3–122. On the other hand, the guarantor is primarily liable upon the instrument upon default by the maker. Because of this, it was concluded in Ligran, Inc. v. Medlawtel, Inc., 174 N.J.Super. 597, 417 A.2d 100 (1980), that the cause of action on a dated demand note accrued against a third party guarantor upon issue rather than upon default by the maker. The nature of the guarantor's liability determines when the cause of action accrues.

D. AGAINST ACCOMMODATION PARTY

Just a quick word here. The accommodation party, as a surety, may sign the instrument in the capacity of maker or indorser. See 3–415(1). Once that capacity is determined, when the cause of action accrues can be determined under the rules in Section 3–122.

E. INTEREST

The parties to a credit instrument typically agree upon a rate of interest to be charged. They may also agree upon a time from which the interest begins to run. If they provide for interest but stipulate neither rate not time, this is construed to mean "interest at the judgment rate at the place of payment from the date of the instrument, or if it is undated from the date of issue." 3–118(d). If there is no provision for interest, the suppletive principles of Section 3–122(4) apply: (1) the rate is that provided by law for a judgment; (2) in the case of parties primarily liable on a demand instrument, interest runs from the date of demand; and (3) in all other cases interest runs from the date of accrual of the cause of action.

CHAPTER 8

PRESENTMENT, DISHONOR, NOTICE OF DISHONOR, AND PROTEST

How a negotiable instrument originates, the ways in which it is transferred from person to person and the rights and duties of the parties on the instrument have been described. It will be recalled that unqualified indorsers and drawers are secondary parties. 3–102(1)(d). As such, they normally are not expected to pay unless the party who is expected to pay (the maker or drawee) fails to do so; and they normally cannot be required to pay unless certain conditions—presentment, dishonor, notice of dishonor, and in some cases, protest—have been satisfied. 3–413(2), 3–414(1), 3–501(1)(a), (b), (c), (2)(a), (b), 3–507(2) and Comment 2, 3–501. In addition to postponing a holder's right to recover from an unqualified indorser or drawer on the basis of secondary liability, a failure or delay in meeting one of these conditions normally has the effect of discharging an unqualified indorser's secondary liability, as well as his liability on the underlying obligation for which he transferred the instrument. 3–502(1)(a), 3–802(1)(b). In limited circumstances a holder's delay or failure in making presentment or in giving notice of dishonor might give the drawer the power to discharge his secondary liability by assigning his rights against the drawee. 3–502(1)(b). In some circumstances a failure or delay in satisfying one of these conditions may give a mark-

Ch. 8   *PRESENTMENT, DISHONOR*

er or acceptor the power to discharge his rights against a bank with whom he has deposited funds. 3–502(1)(b). And a failure to make a required protest may in special circumstances completely discharge the secondary liability of an indorser or drawer. 3–502(2).

The present chapter will consider the legal consequences of failing to satisfy these conditions, the manner in which they are satisfied, and the special circumstances in which the law excuses a failure or delay in satisfying one of these conditions.

§ 1. Effect of Failing to Satisfy Conditions

A. INDORSERS

A holder normally is not entitled to recover from an *unqualified indorser* unless due presentment has been made to the party expected to pay (the maker, drawee or acceptor), 3–501(1)(b), the instrument has been dishonored, the indorser has received due notice of dishonor, and any required protest has been made. In addition, an undue delay with respect to presentment or notice of dishonor or a required protest usually has the effect of *completely discharging* the secondary liability of an unqualified indorser. 3–502(1)(a).

It should be noted, however, that neither presentment, notice of dishonor nor protest is necessary to charge an unqualified indorser whose indorsement includes *words of guarantee* 3–416(5) or an unqualified indorser who indorses *after maturity*. 3–501(4). Consequently, the liability of such parties is not affected by a failure to present, give notice, or make protest.

On the opposite extreme, the *qualified indorser* —one who indorses "without recourse"—is liable only on his warranties and is not liable secondarily. For this reason his liability cannot be affected by a failure or delay with respect to presentment, notice, or protest.

1. Reason for Discharging Secondary Liability of Indorser—Recoupment

If following dishonor an indorser pays the holder, the indorser becomes the owner of the instrument and is entitled to *recoup* —that is, recover what he has paid—from the maker if the instrument is a note or from the drawer or acceptor if the instrument is a draft. In addition, an indorser who pays has a right to recoup from any unqualified indorser prior to himself who has received due notice of dishonor. Each successive indorser who is required to pay is entitled to recoup from the maker, drawer, or acceptor or any prior unqualified indorser who has received due notice of dishonor.

At maturity M delivers a note to P, and by a series of unqualified indorsements, the note is negotiated by P to X to Y to Z. Z duly presents it to M for payment. If the note is dishonored by M, Z has an immediate right to sue M, but this right is likely to have little value because a maker normally pays his notes at maturity if he is able to do so. Fortunately for Z, he is not limited to an action against M. If Z has given due notice of dishonor to Y, X, and P, he is entitled to recover the amount of the note from any of them. If Y pays Z, Y in turn becomes the owner of the note and is entitled to recoup from X who in turn is entitled to recoup from P, whose only rights are against M, the person with whom he dealt.

The right of recoupment obviously can be very valuable. Frequently a valuable right of recoupment can be rendered valueless, however, by a delay in presenting an instrument or in giving notice of dishonor. In collecting money, time frequently is of the essence because payment is made on a "first come, first served" basis. A man who is solvent and willing to pay one day may be insolvent or unwilling to pay the next. There is always the risk that a debtor will waste his fortune, lose it by bad luck, transfer it in fraud of his creditors, or abscond. Undue delay in presenting usually increases the risk that the party expected to pay will not do so. Undue delay in giving notice of dishonor usually increases the risk that any indorser who pays the holder will be unable to recoup what he pays from either the party expected to pay or from a prior indorser.

The law does not permit a holder to saddle this risk unfairly on a prior indorser. Therefore, a holder who, without excuse, delays unduly in presenting an instrument or in giving notice of dishonor is deprived of his right to recover from an indorser on his secondary liability. Since the right of recoupment is given to each successive indorser who may be required to pay, each successive indorser has a similar interest in due presentment and notice of dishonor so that his right of recoupment will not be unduly jeopardized. The law protects his right of recoupment by *completely* discharging him from his secondary liability if there is an undue delay *even when he cannot show that he suffered any loss as the result of the delay.* 3–502(1)(a). Later in this chapter the meaning of *protest* and how protest aids in enforcing the indorser's right of recoupment will be explained. Normally, it is not required, but when it is, a failure to make it on time completely discharges an unqualified indorser even though he cannot show that he suffered any loss as the result of the failure. 3–502(2).

B. DRAWERS

As stated earlier, a drawer normally cannot be required to pay an instrument until certain conditions—presentment, dishonor, and notice of dishonor—have been satisfied. 3–413(2), 3–502(1)(c).

A holder's failure to satisfy these conditions, however, does not, of itself, result in the immediate discharge of the secondary liability of a drawer as it does in the case of the secondary liability of the unqualified indorser. To understand how a failure to satisfy these conditions might bear on the discharge of the liability of a drawer and why it is not given the same effect that it has on the liability of an unqualified indorser it is necessary to keep in mind the difference between the unqualified indorser's and the drawer's relationship to the instrument.

Normally, regardless of whether he receives an instrument as a payee or as a transferee, an unqualified indorser pays fair value for it. When he later negotiates the instrument by his unqualified indorsement, he is unlikely to receive much more than he paid for it;

and in many cases he receives less. Consequently, even if he is completely discharged from liability because a holder fails to make due presentment or to give due notice of dishonor, the unqualified indorser does not receive a windfall. He is merely protected from the possibility that he might not be able to avoid a loss by recouping from some prior party.

The relationship of the drawer to a draft is quite different. It usually costs the drawer virtually nothing to execute and deliver a draft. All he needs is a printed form and a pen. Yet he normally receives consideration in roughly the amount of the draft when he issues it to the payee. Whether he can possibly suffer any loss as the result of a holder's failure to make due presentment or to give due notice of dishonor depends mainly on: a) the arrangements the drawer has made with the drawee, and b) the solvency of the drawee. If the drawee is a total stranger who owes nothing to the drawer the drawer cannot possibly lose anything merely because the holder fails to present the draft for payment or acceptance or, following dishonor, fails to give the drawer due notice of dishonor. If the drawee is indebted to the drawer however, or if the drawer has deposited funds with the drawee for the purpose of paying the draft when it is presented, the failure to make due presentment or to give due notice of dishonor may or may not cause the drawer a loss.

If the drawer suffers no loss as the result of the holder's failure to make due presentment or to give due notice of dishonor, discharging him completely as the result of the failure would grant him a windfall to the extent of whatever he received for issuing the draft. This would be unfair to the holder. But it would be unfair to the drawer to saddle him with a loss if loss did result from the holder's failure to make due presentment or to give due notice of dishonor. With the various possibilities in mind, the framers of the Code determined to give the drawer a right to obtain a discharge from liability, but only to the extent necessary to enable him to avoid any losses he might otherwise suffer as the result of the holder's failure to make due presentment or to give due notice of dishonor. The extent of this right and the manner of exercising it are provided in Section 3–502(1)(b) which states:

> Where without excuse any necessary presentment or notice of dishonor is delayed beyond the time when it is due * * * any drawer * * * who because the drawee * * * becomes insolvent during the delay is deprived of funds maintained with the drawee * * * to cover the instrument may discharge his liability by written assignment to the holder of his rights against the drawee * * * in respect of such funds, but such drawer * * * is not otherwise discharged.

The following examples illustrate how a failure to make proper presentment or to give due notice of dishonor affects a drawer's liability under the Code.

Example 33

On April 1st, R signs and delivers to P, payee, a draft for $5,000 payable on July 1st, drawn on E. On May 1st P indorses "P" and delivers the draft to H for value. H fails to present the draft to E for payment until August 1st at which time E flatly refuses to pay although he admits that R has deposited $5,000 with him to meet the draft when it fell due and that he, E, is completely solvent and capable of paying the draft. H promptly notifies R of E's refusal and demands payment from R. R refuses to pay, contending that he has been discharged by H's delay in presenting the draft to E for payment. H is entitled to recover from R the full amount of the draft. A drawer is not discharged from his secondary liability automatically upon a holder's failure to make due presentment or to give due notice of dishonor. Moreover, since E did not become insolvent during H's delay in making presentment, R cannot obtain his discharge by making an assignment to H of his rights against E with respect to the funds on deposit. In short, R will have to pay H and then proceed against E for breach of the contract of deposit. In contrast, as shown earlier, P, who indorsed without qualification, is completely and automatically discharged from his secondary liability even though P suffered no loss as the result of the delay in presentment.

Example 34

On May 1st R places $10,000 in the hands of E to meet a draft for that sum. On May 15th R issues the draft to P. The draft is payable on June 1st. On this latter date, E is solvent and able to pay the draft, but P does not present the draft for payment until October 1st at which time E is insolvent and able to pay creditors only thirty cents on a dollar. Although R is not automatically discharged by the delay in presentment, he can obtain a discharge by giving P a written assignment of his claim for $10,000 against E. Since P is responsible for the unfortunate situation, he is required to suffer the loss and the inconvenience of following E's insolvency proceeding.

When protest is required, a failure to make it completely discharges a drawer just as it completely discharges an unqualified indorser. 3–502(2).

C. DRAWEES

Since a drawee as such is not liable on a draft but only on his contract with the drawer, his position is quite different from that of either an unqualified indorser or a drawer; and his liability is not affected at all by a failure to present or to give due notice of dishonor in the manner required to activate or retain the liability of these parties.

D. MAKERS AND ACCEPTORS

Since makers and acceptors are liable *primarily* rather than *secondarily*, they normally are required to pay at maturity and have no right of recoupment against anyone else if they are required to pay. Therefore, the liability of a maker or acceptor normally is not affected by a holder's failure to make due presentment or to give due notice of dishonor. If, for example, the promise is to pay at a stated time, the cause of action accrues "on the day after maturity," 3–122(1)(a), and may be sued upon without presentment until the statute of limitation runs. See Chapter 7, Section 5.

Sometimes, however, a maker or acceptor deposits funds and arranges for someone else to make the actual payment on his behalf. When he does, the position of the maker or acceptor is quite similar to that of the drawer who deposits funds with a drawee to pay a draft. If the depositary refuses to pay as agreed, the maker or acceptor has a right of recoupment from the depositary. If the depositary happens to be a *bank*— but not otherwise—the Code seeks to protect the mak-

Ch. 8　　*PRESENTMENT, DISHONOR*

er's or acceptor's right of recoupment against any loss that might result from the bank's becoming insolvent during the holder's delay in making presentment or in giving notice. It does this in the same way it protects the drawer's right of recoupment in similar circumstances—by granting the maker or acceptor a discharge from liability upon his making a written assignment to the holder of his rights against the payor bank with respect to such funds. 3–502(1)(b). The liability of a maker or acceptor is not affected by the holder's failure to make protest.

In summary, drawers and unqualified indorsers are secondary parties on an instrument. 3–102(1)(d). Both are conditionally liable, 3–413(2) and 3–414(1) and one of those conditions is presentment, that is, a demand by the holder upon the maker for payment. 3–504(1). See 3–501(1)(b) and (c). But differences in the risk and opportunities for recoupment justify a different treatment of indorsers and drawers when an unexcused failure properly to present occurs: an indorser is automatically discharged, 3–502(1)(a) but a drawer must satisfy the restrictions of 3–502(1)(b). In either case, however, the failure of a holder to make a proper presentment, although not affecting the maker's primary liability, impairs the liability of secondary parties. To this extent, therefore, proper presentment is an essential step to full protection of the holder's rights on the instrument.

§ 2. Satisfying Requirements

In most cases the handling of matters relating to presentment, notice of dishonor and protest is left to banks whose specialized personnel are familiar with the requirements and have established routines for satisfying them. There remain, however, a substantial number of businesses who find it worthwhile to handle these matters themselves.

A. PRESENTMENT

Unless presentment is excused, there can be no dishonor of an instrument which is not duly presented; and if there is no dishonor, there is no right to proceed against drawers and indorsers on their secondary liability.

A presentment may be made either for the purpose of obtaining *acceptance* of a draft or check or for the purpose of obtaining *payment* of any kind of instrument. As will be explained, the principles that determine in which cases presentment is required differ in some respects for the two types of presentment. But the legal effect of an unexcused failure or delay in making a necessary presentment is the same for both; and the legal requirements for making an effective presentment are substantially the same for both.

1. WHAT CONSTITUTES EFFECTIVE PRESENTMENT

Basically, presentment for acceptance is a *demand* for acceptance and presentment for payment is a *demand* for payment. 3–504(1). To be effective as a presentment, however, the demand may have to satisfy certain requirements. What these requirements are depends in large part on what, if anything, is requested at the time by the person upon whom the demand is made. According to Official Comment 1 to Section 3–504, if nothing more is requested by the person upon whom the demand is made, any demand upon the party to pay, no matter where or how made, is an effective presentment. Of course, it must be made by or on behalf of the holder and it must be made upon the maker, acceptor, drawee, or other payor. Presentment may be, and frequently is, made through a clearing house. 3–504(2)(b). It may be made by mail in which case it is effective when the letter is received. 3–504(2)(a). Although Comment 1, mentioned above, seems to imply that presentment may be made over the telephone, there is a case to the contrary. See Kirby v. Bergfield, 186 Neb. 242, 182 N.W.2d 205 (1970). Even assuming that an effective presentment can be made by telephone, however, it is usually desirable in case of dishonor by telephone to make a second demand by another means rather than rely on the telephone conversation. Normally, presentment is made at the place of business or residence of the party who is expected to pay or accept, but unless some other place is provided in the instrument or objection is raised immediately, the Code allows the holder to make an effective presentment wherever he can find the payor or someone authorized to act for him. Comment 1, 3–504.

To balance the liberal attitude concerning what might suffice as a presentment in the first instance,

Section 3-505(1) gives the party on whom presentment is made the power to require
  1. exhibition of the instrument;
  2. reasonable identification of the person making presentment;
  3. evidence of authority if presentment is made for another;
  4. production of the instrument at a place specified in it or, if none is specified, at any reasonable place; and
  5. a signed receipt on the instrument for any partial or full payment and its surrender upon full payment.

If the party on whom presentment is made requires any of these things, the presentment is ineffective unless the presenting party complies within a reasonable time. 3-505(2). If the instrument is a draft accepted or a note made payable at a bank in the United States, the instrument must be presented at such bank even though the party on whom presentment is made does not require it. 3-504(4). If the instrument has two or more makers, acceptors, or drawees; an effective demand may be made on any one of them. 3-504(3)(a).

Example 35

On May 1, D drew a time draft for $5,000 on E, payable on June 1, and issued it to P. On June 1 at 2 PM, P telephoned E at her place of business, identified himself and demanded payment of the draft. P, who was in San Francisco, directed E, who was in Denver, to wire the money to his bank. Although the demand was timely, 3-503(1)(b), it is doubtful that a demand over the telephone is a proper presentment under 3-504(2). Until a presentment is made, the draft cannot be dishonored and the drawer is not liable on the instrument. Even if the presentment was proper, E may "without dishonor" require P to exhibit the instrument. 3–505(1)(a). P's inability promptly to comply would invalidate the presentment. But if P mailed the instrument with a demand for payment within a reasonable time, the presentment would be proper and E's time for payment would run "from the time of compliance." 3–505(2).

2. Time of Presentment

The time of presentment involves the hour of the day and the day. The first presents few problems as presentment normally must be made at a reasonable hour, and if made at a bank must be made during its banking day. 3–503(4). Determining the *day* of presentment is the most critical problem faced by a holder intending to satisfy the requirements regarding presentment. This decision may depend on whether the presentment is made to obtain payment or to obtain acceptance, the nature of the instrument, the position of the secondary party whose liability the holder wishes to fix, and other factors.

a. Presentment for Acceptance

There are only three situations in which presentment for acceptance is required to fix or retain the secondary liability of indorsers or drawers. The *first* occurs when the draft itself provides that it is to be presented for acceptance. The *second* occurs when the draft is made payable elsewhere than at the drawee's residence or place of business. The *third* occurs when the date of payment depends upon presentment for acceptance, as when a draft is payable a stated period after sight. 3–501(1)(a).

In these three situations the day when presentment for acceptance must be made depends on whether it is payable (1) at or a fixed period after a stated date, (2) after sight, or (3) on demand. If payable at or a fixed period after a stated date, such presentment must be made on or before the date it is payable. 3–503(1)(a). If payable after sight, it must be presented for acceptance *or negotiated* within a reasonable time after date or issue, whichever is later. 3–503(1)(b). If payable on demand, a necessary presentment for acceptance must be made within a reasonable time after the secondary party whose liability is under consideration became liable on the instrument. 3–503(1)(e). Normally a drawer becomes liable when he issues a draft, and an indorser becomes liable when he indorses and delivers it.

Even though presentment for acceptance is not required, a holder may present a draft for acceptance. 3–501(1)(a). Unless it is payable at a stated date, however, a drawee's refusal to accept a draft which the holder is not required to present for acceptance is not a dishonor. If a draft is payable at a stated date, optional presentment for acceptance must be made within the same time as if presentment for acceptance is required—i. e., on or before the date stated.

b. *Presentment for Payment*

If an instrument *shows the date on which it is payable,* presentment for payment is due on that date. 3–503(1)(c). This means that a primary party can be required to pay on that date and that a refusal to pay by a maker, acceptor or drawee following a due demand for payment on that date constitutes a dishonor. It also means that presentment on that date is necessary to fix or retain the liability of parties who are entitled to insist on presentment. An instrument is deemed to show the date on which it is payable if it states that it is payable: (1) on a specific date, (2) a fixed period after a specific date, or (3) a fixed period after sight and the acceptance indicates the date of acceptance. In the two latter situations, the time for presentment for payment is determined by excluding the day from which the time is to begin and including the date of payment. For example, if an instrument is dated June 1st and is payable "twenty days after date," it becomes due on June 21st. (If it is not paid on that date the holder is entitled to sue on the following day. 3–122(1)(a)). This principle applies also when calculating the time for making presentment for acceptance. In fact, this method of calculation is used in determining time throughout the law of contracts and is not limited to the law of negotiable instruments. Comment 1, 3–503. If a presentment falls on a day which is not a full business day for the person making the presentment or which is not a full business day for the person who is expected to pay or accept, presentment is due on the next following day that is a full business day for both parties. 3–503(3).

The proper time for presenting a *demand* instrument for payment depends on the purpose for which presentment is made. If the purpose is to *render a refusal to pay a dishonor* so as to entitle the holder to sue a secondary party immediately, the presentment may be made upon its date, or if no date is stated, on the date of issue. 3–122(1)(b). If the purpose is to *retain the liability of a secondary party*, presentment is timely if it is made within a reasonable time after the secondary party became liable on the instrument. 3–503(1)(e). To determine whether an *indorser* is discharged because of a delay in presenting a demand instrument for payment, a reasonable time is measured from the time he negotiates the instrument. To determine whether a presentment is effective with respect to a *drawer*, a reasonable time is measured from the time of issue. To determine whether a *maker* or *acceptor* with funds on deposit in a bank may obtain a discharge by assigning his rights to the holder, presentment for payment must be made within a reasonable time after issue, in the case of a maker, and within a reasonable time after acceptance, in the case of an acceptor. 3–503(1)(e).

c. Reasonable Time for Presentment

If the issue is raised in a case in which presentment must be made "within a reasonable time," the party seeking to recover has the burden of proving that the demand for payment or acceptance was made within a reasonable time. What constitutes a reasonable time to present is determined by the nature of the instrument, relevant trade and banking usage, and the facts of the particular case. 3–503(2). In deciding this question, courts may consider a variety of circumstances such as the distances involved, communications and transportation facilities available, whether the instrument is intended to serve as continuing security, and perhaps of most importance, whether the instrument bears interest. It is impossible to reconcile the numerous cases dealing with this problem.

In the case of an ordinary uncertified check drawn and payable within the United States, it is *presumed* that a reasonable length of time in which to present for payment or initiate bank collection is seven days after indorsement to determine whether an *indorser's* liability is discharged, or thirty days after date or issue, whichever is later, to determine whether a *drawer* may discharge his liability by making an assignment of his rights against the bank. 3–503(2). *These presumptions are not conclusive.* They may be rebutted by competent evidence material to the question of reasonableness in the case under consideration. For example, if a check is drawn for a large sum and the drawer, drawee, and payee all are situated in the same city, a court is likely to consider it to be unreasonable to delay thirty days before presenting the check for payment.

If an instrument that is otherwise payable at a definite time contains an acceleration clause, and the event accelerating the time for payment occurs, presentment for payment is due within a reasonable time thereafter. 3–503(1)(d). In addition to the factors that would normally be taken into consideration in determining a reasonable time, it would be necessary here to take into account the knowledge of the fact of acceleration which the holder, or other person charged with responsibility for making the presentment, might or might not have.

NOTICE OF DISHONOR & PROTEST Ch. 8

d. *Presentment by Collecting Bank*

In general, banks are governed by the principles described above. There is one important exception. Usually, if an instrument shows the date on which it is payable, presentment for payment must be made *on that date*. 3–503(1)(c). However, in the case of an instrument that is *not* payable by, through or at a bank, a collecting bank normally can make a proper presentment for payment of such an instrument by sending notice that it holds such item for collection in time to be received *on or before* the day when presentment is due. 4–210(1).

In Batchelder v. Granite Trust Co., 339 Mass. 20, 157 N.E.2d 540 (1959), it was held that presentment was sufficient to retain an indorser's secondary liability on a note when the bank, ten days before the note fell due, mailed notice to the maker that it held the note and then retained the note until it was due. If a proper request—for example, exhibition of the instrument or evidence of authority to present—is made by the party expected to accept or pay, however, the bank must comply with the request by the end of the bank's next banking day after it learns of the request. 4–210(1).

e. *Domiciled Instrument*

Often the maker of a note or the acceptor of a draft makes it payable at a particular place or bank. Such notes or drafts are referred to as *domiciled* instruments. Sometimes the place of payment is indicated only for convenience. At other times it is intended to prevent the holder of an instrument bearing a favorable interest rate and *payable at a stated time* from taking advantage of the favorable interest rate by extending the time simply by not presenting it for payment until shortly before the statute of limitations has run. Domiciled instruments are effective for this purpose because a tender of payment to the holder of any instrument when or after it is due discharges the person making the tender to the extent of all subsequent liability for interest, costs and attorney's fees 3–604(1); and the fact that the maker or acceptor of an instrument payable *otherwise than on demand* is able and ready to pay at every place of payment specified in the instrument when it is due, is equivalent to tender. 3–604(3).

Since the maker of a *demand* instrument must make an *actual* tender of full payment in order to obtain a discharge to the extent of subsequent interest, costs and attorney's fees, he cannot enjoy the above major advantage that accrues to one issuing a domiciled instrument that is payable at a *definite* time. It appears, however, that he can effectively offer the payee the principal advantage of a demand instrument —the right to demand payment at any time—without himself incurring the risk of being required to pay interest until shortly before the statute of limitations has run out. He can accomplish this by issuing a domiciled instrument for some fixed period satisfactory to himself and including a provision giving the holder the right to accelerate payment by demanding payment at any earlier time.

Under the Code, the terms of the draft are held not to be varied by an acceptance that provides for payment at a bank or place in the United States unless the acceptance provides that the draft is to be paid *only* at such bank or place. 3–412(2). Whether a note or acceptance stating that it is payable at a bank is equivalent to a draft drawn on the bank payable when it falls due out of any funds of the maker or acceptor in current account or otherwise available for such payment depends on which of two alternative provisions of the Code is adopted. 3–121.

B. DISHONOR

1. WHAT CONSTITUTES

In general, an instrument is dishonored if a *necessary* or *optional* presentment is duly made and due acceptance or payment is refused or cannot be obtained within the time prescribed. 3–507(1)(a).

Presentment for *payment* is always *necessary* unless excused. 3–501(1)(b), (c). As stated above, presentment for *acceptance* is *necessary* in only three situations. (See pages 188–189). If presentment for *acceptance* is not *necessary*, it is *optional* if, but only if, the draft is payable *at a stated date*. 3–501(1)(a). Presentment for *acceptance* of a *demand* draft is never *optional* although it may be *necessary* in one of the three circumstances referred to above. Therefore, if presentment for acceptance of a demand draft is not necessary, a refusal to accept it does not constitute a dishonor. For example, a bank's refusal to certify an ordinary check for a holder who would prefer the bank's liability to cash does not constitute a dishonor. Consequently, the holder would not, by giving notice of such refusal, acquire an immediate right to recover from either a drawer or an unqualified indorser on the basis of secondary liability.

If a party upon whom a proper demand for acceptance or payment has been made flatly refuses to comply or refuses to comply unless the presenting party does more than legally required, the dishonor is clear. Dishonor is equally clear if the drawee of a draft announces that the drawer has insufficient funds on deposit or has stopped payment. In contrast, dishonor does not occur if the party on whom the demand is made refuses to comply until the holder has satisfied some legal requirement. For example, as previously mentioned, the party to whom presentment is made may insist that the holder exhibit the instrument, that he produce it at a place specified in it, or that a necessary indorsement be obtained. Comment 2, 3–510.

Similarly, it does not constitute dishonor if the person on whom a demand for *payment* has been made defers payment pending a reasonable examination to determine whether the instrument is properly payable, as long as payment is made before the close of business on the day the demand is made. 3–506(2). A person upon whom a demand for *acceptance* has been made may, without dishonor, defer acceptance until the close of the next business day following presentment. 3–506(1).

In any case in which a necessary presentment is excused (see pages 203–205), dishonor occurs if the instrument is not duly accepted or paid. 3–507(1)(b).

The foregoing principles are modified in some respects where the presentment is made by or on behalf of a bank.

C. NOTICE OF DISHONOR

If, following dishonor, due notice of dishonor is not given, the effect is the same as if there had been undue delay in making presentment. However, if a necessary notice is given, the holder has an immediate right to recover from the party to whom it is given. 3–507(2).

1. METHOD OF GIVING NOTICE

Notice of dishonor may be given in any reasonable manner. 3–508(3). For example, it may be given orally face to face or over the telelphone, by mail, by telegraph, or through a clearing house. However, the person giving notice should remember when choosing a means of communication that he may be required to prove in court that he has given notice. Need for proof is one reason why notice of dishonor usually is given in writing and why, if it is given orally, it usually is confirmed in writing. It is noteworthy that a written notice of dishonor is effective when sent even though it is not received by the addressee. 3–508(4). Accordingly notice sent by ordinary mail is given as soon as a properly addressed letter bearing sufficient postage and containing the notice is deposited in a mail box, chute, or other receptacle maintained by the Post Office for mailing letters.

No special words must be used to give notice of dishonor. Sending the dishonored instrument with a stamp, ticket, or writing, stating that acceptance or payment has been refused is sufficient. Accordingly a certificate of protest or a notice of debit with respect to the instrument is sufficient. 3–508(3). Although the notice is intended to convey the idea that the party giving notice is asserting his rights against the party given notice, this need not be expressly stated as it is clearly implied by the notice. Any words that identify the instrument and state that it has been dishonored are sufficient. 3–508(3).

2. Persons Given Notice

Notice of dishonor may be given to any person who may be liable on the instrument. 3–508(1). Notice to one partner is notice to every other partner. 3–508(5). Notice to joint parties who are not partners must be given to each of them unless one is authorized to receive such notice for the others. When a party is in insolvency proceedings instituted after the issue of the instrument, notice may be given either to him or to the representative of his estate, for example, a trustee in bankruptcy or an assignee for the benefit of creditors. 3–508(6). When a person entitled to notice is dead or incompetent, notice may be sent to his last known address or given to his personal representative. 3–508(7).

3. Persons Giving Notice

Notice of dishonor may be given by or on behalf of the holder or any party who has received notice or can be compelled to pay the instrument. In addition, an agent or bank in whose hands the instrument is dishonored may give notice to the principal or customer or to another agent or bank from which the instrument was received. 3–508(1).

4. Sequence in Which Notice May Be Given

Since notice of dishonor is intended to facilitate the recoupment by each party required to pay, it need not be given in any special order. Often a holder gives notice of dishonor only to his immediate transferor, assuming that each successive party receiving notice will in turn give notice to his transferor. If the holder's transferor pays, this procedure is sufficient for the holder. But since it may be impossible to collect from his immediate transferor, it is advisable for the holder to give notice to all the prior indorsers he can locate so that if he cannot recoup from one he can from another. When duly given, notice of dishonor operates for the benefit of all parties who have rights on the instrument against the person notified. 3–508(8).

5. Time Allowed For Giving Notice

When given by the holder at dishonor to his transferor, notice usually must be given before midnight of the third business day after dishonor. When given by any prior holder, it must be given before midnight of the third business day after that party receives notice of dishonor. 3–508(2). However, a bank is required to give notice before its midnight deadline, which is midnight of the banking day following the banking day on which the bank receives the item or notice of dishonor. 3–508(2), 4–104(1)(h).

D. PROTEST

The term *protest* has several meanings. Often it refers to the entire procedure of presenting, giving notice of dishonor, and protesting. Sometimes it refers only to the making of protest.

Used in the technical sense, protest refers to a certificate of dishonor made under the hand and seal of a United States consul or vice consul, a notary public or another person authorized to certify dishonor under the law of the place where it occurs. 3–509(1). In practice, protest is almost always made by a notary public. Although it need not be in any special form, it "must identify the instrument and certify either that due presentment has been made or the reason why it is excused and that the instrument has been dishonored by nonacceptance or nonpayment." 3–509(2).

Under the prior law, the notary public was required to have actual knowledge of the essential facts which he certified. Under the Code it is sufficient if the protest is made on information satisfactory to him. 3–509(1). Typically, satisfactory information might consist of information furnished by the party making presentment, the admission of the dishonoring party, or the fact that the instrument was not paid when it was presented a second time. Comment 4, 3–509. It is assumed that there usually is no sound motive for making a false protest; consequently, the basis on which protest is made is rarely questioned. Comment 4, 3–509.

1. WHEN PROTEST IS REQUIRED

Protest is required only when there has been a dishonor of a draft which appears on its face to be either drawn or payable outside of the states, territories, dependencies and possessions of the United States, the District of Columbia and the Commonwealth of Puerto Rico. 3–501(3). Such a draft is commonly referred to as a *foreign bill*. An unexcused failure to protest such an instrument following dishonor completely discharges the drawer as well as all indorsers whether or not they have incurred any loss from the failure. 3–502(2). This result contrasts with the legal effect of the failure to make due presentment or to give due notice of dishonor which completely discharges indorsers but gives a drawer, at most, a right to obtain a discharge by making an assignment of his rights against the drawee. Even though a dishonored instrument is not a foreign bill and protest is not required, it may be, and very frequently is, protested because of the advantages of protest in proving dishonor.

2. TIME FOR MAKING PROTEST

Any necessary protest is due within the same time that notice of dishonor is due. 3–509(4). Any protest that is not necessary may be made at any time before it is offered in evidence. Comment 7, 3–509.

3. PROTEST AS EVIDENCE

Protest, as a requirement, is intended to protect the right of recoupment of drawers and indorsers of foreign bills. Since foreign bills are made in one country and payable in another, proving dishonor of such instruments may be difficult. Protest reduces this difficulty substantially because a certificate of protest is admissible as evidence and creates a rebuttable presumption of dishonor. This is one reason why it is customary to protest even those instruments which are not required to be protested. Also, it is common practice to include in the certificate of protest a certification that notice of dishonor has been given. Comment 6, 3–509. The inclusion of this provision in the certificate of protest creates a rebuttable presumption that notice of dishonor has been given as stated. 3–510(a).

4. FORWARDING PROTEST

It is not necessary in order to satisfy the requirement of protest to forward a copy of the protest to persons who are entitled to insist on protest. Since it serves as a notice of dishonor, however, it is customary, when protest is made, to send a copy of the certificate of protest to all persons who are entitled to notice of dishonor. This is so regardless of whether or not protest was required. The original certificate of protest usually is attached to the dishonored instrument and goes with the instrument to each person who pays and takes up the instrument.

5. OTHER TYPES OF EVIDENCE

Although a certificate of protest has been the most frequently offered evidence of dishonor and notice of dishonor, there are other types of evidence which may be admitted. For example, oral testimony by the party presenting the instrument and giving the notice may be admitted. Likewise, entries made in the regular course of business sometimes may be admitted. Before the Code, protest was the only type of evidence which had the advantage of creating a rebuttable presumption in favor of the facts it purported to establish. Under the Code, there is also a rebuttable presumption of any dishonor or notice of dishonor which is shown in (1) a purported stamp or writing of the drawee, payor bank, or presenting bank on the instrument or accompanying it which states that acceptance or payment has been refused for reasons consistent with dishonor; or (2) any book or record of the drawee, payor bank, or collecting bank kept in the usual course of business which shows dishonor even though there is no evidence to indicate who made the entry. 3–510(b), (c).

§ 3. Excusing Omission or Delay in Presentment, Notice or Protest

In some cases, making proper presentment, giving due notice, and making proper protest would be an empty formality contributing nothing to the persons intended to benefit by them. In other cases, whatever advantage might accrue from complying with these requirements is more than outweighed by the disadvantage to the person of whom compliance normally is required. Consequently, the law makes allowance by excusing either a delay or omission in a number of circumstances as described in Section 3–511 of the Code which provides:

(1) Delay in presentment, protest or notice of dishonor is excused when the party is without notice that it is due or when the delay is caused by circumstances beyond his control and he exercises reasonable diligence after the cause of the delay ceases to operate.

(2) Presentment or notice or protest as the case may be is entirely excused when
  (a) the party to be charged has waived it expressly or by implication either before or after it is due; or
  (b) such party has himself dishonored the instrument or has countermanded payment or otherwise has no reason to expect or right to require that the instrument be accepted or paid; or
  (c) by reasonable diligence the presentment or protest cannot be made or the notice given.

(3) Presentment is also entirely excused when
  (a) the maker, acceptor or drawee of any instrument except a documentary draft is dead or in insolvency proceedings instituted after the issue of the instrument; or
  (b) acceptance or payment is refused but not for want of proper presentment.

(4) Where a draft has been dishonored by nonacceptance a later presentment for payment and any notice of dishonor and protest for nonpayment are excused unless in the meantime the instrument has been accepted.

(5) A waiver of protest is also a waiver of presentment and of notice of dishonor even though protest is not required.

(6) Where a waiver of presentment or notice of protest is embodied in the instrument itself it is binding upon all parties; but where it is written above the signature of an indorser it binds him only.

Presentment for acceptance or payment also is completely excused if (1) the place of acceptance or payment is specified in the instrument and neither the party to accept or pay nor anyone authorized to act for him is accessible at such place or (2) the place of payment is *not* specified and neither the party to accept or pay nor anyone authorized to act for him is accessible at his place of business or residence. 3–504(2)(c).

§ 4. Effect of Excusing Omission or Delay

An excuse may extend to one or more conditions and may affect the liability of one or more parties. Regardless of which conditions and whose liabilities are affected, the legal effect of excusing an omission or delay is the same as if each condition excused had been met within the time normally allowed.

Example 36

(This example is based upon Clements v. Central Bank of Georgia, 155 Ga.App. 27, 270 S.E.2d 194 (1980).)

To pay an existing debt, Drawer drew a draft in the amount of $30,000 against his "money market" account with Drawee to the order of Payee. P, by an unqualified indorsement, negotiated the draft to Holder, a brokerage house, as payment for a margin call. On June 5, H mailed the draft to Drawee and it was received on the morning of June 7. On June 7, H heard that Drawer was having financial problems and, at 11 AM, telephoned Drawee to determine if the draft would be paid. An officer of Drawee stated that the status of Drawer's account was "confused" and to "call back later in the day." At 5 PM that afternoon, H telephoned again and was told that the draft "would not be paid that day." At Drawee's suggestion, H agreed to leave the draft there with the expectation that it would be paid when there were sufficient funds. On June 10, H telephoned again and was told that the draft would not be paid—Drawer had closed his account. H requested that the draft be returned and it was received on June 12. At 4 PM on June 12, H telegraphed P that the draft had been dishonored and, in due course, demanded payment. P refused on the ground that H's delay in giving notice of dishonor had discharged him from liability on the instrument.

P is correct. Unless excused under Section 3–511, P was entitled to a notice of dishonor. 3–414(1) and 3–501(2)(a). Unless excused, an indorser is discharged when any necessary notice is "delayed beyond the time it is due." 3–502(1)(a). When the party seeking to enforce the instrument is not a bank (this case), the necessary notice must be given before midnight of the third business day "after dishonor or receipt of notice of dishonor." The questions are, therefore, did Drawee dishonor, if so when, and when did H learn of it? Under Section 3–506(2), Drawee could defer payment to the close of the business day of receipt without dishonor. But the instrument is dishonored if payment "cannot be obtained within the prescribed time * * *" 3–507(1)(a). Drawee, then, dishonored on June 7 and H had notice of that dishonor on that date. See 3–508(3). Thus, H should have given P notice of the dishonor not later than midnight of June 10. The June 12 notice was too late and, on these facts, was not excused under Section 3–511(1). Accordingly, P, an indorser, is discharged from liability on the instrument. 3–601(1)(i).

CHAPTER 9

LIABILITY BASED ON WARRANTIES

Whether or not a person signs an instrument, he may be liable on the basis of certain warranties. These warranties fall into two categories: (1) those imposed upon persons who *transfer* instruments, 3–417(2); and (2) those imposed on persons who *obtain payment or acceptance*, 3–417(1). The same scope and classification also obtains in the check collection process, governed primarily by Article 4. See 4–207.

Unless disclaimed by agreement, warranties in Articles 3 and 4 allocate specialized risks in dealing with commercial paper. Compare Sections 2–312 through 2–315 of Article 2, which impose warranties of quality in the sale of goods. They should also be contrasted with the more limited warranties made by an assignor of contract rights. See Chapter 3, Section 1A.

§ 1. Transferors' Warranties

In the absence of a contrary agreement, any person who transfers an instrument and receives consideration gives five separate implied warranties relating to (1) title, (2) signatures, (3) alterations, (4) defenses, and (5) insolvency proceedings. 3–417(2). These warranties arise regardless of whether the transfer is by assignment or is by negotiation, with or without indorsement. With one minor exception which will be mentioned, they are the same for all transferors. Transferors by indorsement, qualified as well as unqualified, give these warranties *not only to their immediate transferees but also to any subsequent holders who take the instrument in good faith.* Unless he is a depositor for collection or a collecting bank, however, one who transfers an instrument without indorsement gives these warranties only to his immediate transferee. 3–417(2), 4–207(3).

Confusion is sometimes caused by the fact that in addition to giving the five transferors' warranties mentioned above, transferors give the same three *presenters'* warranties that are given by those who obtain payment or acceptance. 3–417(1). Although, as will become clear, there are several important differences between the two classes of warranties, it should be sufficient here to mention the basic difference: presenters' warranties run only to those who pay or accept whereas transferors' warranties run only to the transferor's immediate transferee and later holders.

A. WHY HOLDER MAY PREFER TO SUE TRANSFEROR ON WARRANTY

Because transferors of negotiable instruments usually indorse without qualification, secondary liability normally furnishes a sufficient basis for recovering from a transferor when the party who is expected to pay does not. But if the holder's right to recover on the basis of secondary liability is barred because the instrument was transferred without an indorsement or by a qualified indorsement or if the secondary liability of an unqualified indorser is discharged by the holder's failure to satisfy the requirements with respect to presentment, dishonor, notice of dishonor, or protest, the holder often may still avoid or minimize his loss by suing a prior party on one or more of the transferors' warranties. Even though the defendant indorsed without qualification and his secondary liability has not been discharged, it may be desirable for the holder to sue on the basis of a breach of warranty, rather than on secondary liability, because his right to sue for breach of warranty arises immediately on discovering the breach though the time for payment and possible dishonor has not arrived. In cases where the aggrieved party is entitled to sue on an underlying obligation, he may sometimes prefer to proceed on the basis of the breach of warranty because it is easier to prove. Finally, proceeding on the basis of breach of warranty allows the aggrieved party to rescind rather than seek money damages.

B. WARRANTY OF TITLE

The first warranty given by the transferor for consideration is that "he has good title to the instrument or is authorized to obtain payment or acceptance on behalf of one who has a good title and that the transfer is otherwise rightful." 3–417(2)(a). This warranty is intended to protect the transferee against the risk that the transferor had neither title nor proper authority and that as a result: (1) the true owner, who might be either the person for whom the transferor purported to act or some other person, might assert his claim and bring an action to recover the instrument or its value; (2) the transferee's own lack of title might be asserted against him as a defense when he demands payment; (3) the transferee will be liable for breach of the same warranty if he later sells the instrument; or (4) if the transferee or some later party obtains payment or acceptance from the party who is expected to pay, he may be held liable for a breach of the *presenters'* warranty of title which will be considered shortly. (See pages 222–225) Compare UCC 2–312.

A transferor might breach the warranty of title in any of several ways. The signature of the payee or some other party in the chain of title might have been forged by the transferor or another. A finder, thief, or unauthorized agent, might sell an instrument that is payable to someone else who has not indorsed it. In cases such as these, the transferor is not a holder or an owner, and neither is his transferee. The warranty is obviously broken.

In some cases, however, the warranty is broken even though the transferor is actually the holder. Thus, an instrument that is payable to bearer or to the order of the transferor might have been found or stolen by him or acquired by him as agent and transferred without authority of the owner. In cases of this kind, although the transferor is a holder because the instrument runs to him, he breaks the warranty of title because he is not the owner and does not have authority from the owner. In short, the transfer is not "otherwise rightful". In these cases, if the transferee acquires the instrument as a holder in due course, however, the latter gets good title because a holder in due course gets good title by negotiation from even a finder, a thief, or an unauthorized agent. Nonetheless, the warranty of title is broken.

When a breach of this warranty results from a forgery in the chain of title, there can be no later holder or holder in due course unless the person whose signature is forged reacquires the instrument. Normally, therefore, if there is a forgery in the chain of title, this warranty is broken not only by the forger but also by any later transferor.

> *Example 37*
>
> M delivers his note to P, payee, as payment of a debt. T steals the note, forges P's blank unqualified indorsement, and sells the note to A. A sells the note to B without indorsing. B indorses "without recourse" and sells the note to C. C indorses without qualification and sells the note to H. At maturity M refuses to pay H. Since the note runs to P, who never indorsed, T, A, B, and C broke the transferor's warranty of title. As an unqualified indorser, C is liable to H secondarily as well as on the warranty. The same is true of T whose forged indorsement operates as if he had signed his own name. As a qualified indorser, B is not liable secondarily but, as an indorser, his warranty liability runs through to H. As a transferor without indorsement, A is not liable secondarily, and his liability on the warranty runs only to his immediate transferee, B, not to C or H.

C. WARRANTY SIGNATURES ARE GENUINE OR AUTHORIZED

The second warranty given by a transferor for consideration is that "all signatures are genuine or authorized." 3–417(2)(b). If a signature of a payee or an indorsee in the chain of title is unauthorized, there is a breach of this warranty as well as a breach of the warranty of title. If the signature of a maker, drawer, drawee, acceptor, or indorsee not in the chain of title is unauthorized, however, there is a breach of this warranty but no breach of the warranty of title.

The most obvious purpose of this warranty is to protect the transferee against the risk that he will not be able to collect because some person who appears to be liable on the instrument is not. The warranty also protects the transferee against the further risk that he will be liable on the same warranty if he transfers the instrument. Also, depending on the facts, if he or his transferee acquires payment or acceptance after learning that the signature of the maker or drawer is unauthorized, he might be held liable for breach of the presenters' warranty that he has no such knowledge. (See pages 225–229)

Example 38

P, an employee of M, obtains a blank note from a bank, fills it in "to the order of P" and forges M's signature as maker. P negotiates the note to A for consideration. A then negotiates the note to H for consideration "without recourse." Five months before the note is due, H learns of the forgery. To the surprise of no one, P has disappeared. What should H do?

If H should wait until the note was due and dishonored by M, an action on the note against M or A would not succeed. M's signature was "wholly inoperative, 3–404(1), and A indorsed "without recourse." 3–414(1). Further, since the note was never issued, H cannot be a holder and there was never an underlying obligation to be suspended. H, then, must sue A for breach of warranty under 3–417(2). But which warranty? The best bet is the warranty that "all signa-

Ch. 9 *LIABILITY BASED ON WARRANTY*

tures are genuine or authorized." 3–417(2)(b). A, by transferring the note, warranted that M's signature as maker was genuine. A, however, did not breach the warranty of good title. Except for disputes over "rightful" transfer, it has been held that the warranty is limited to invalid signatures in the chain of necessary indorsements. Sun 'N Sand, Inc. v. United California Bank, 21 Cal.3d 671, 148 Cal.Rptr. 329, 582 P.2d 920 (1978).

Example 39

X, an employee of P, stole a check issued by M "to the order of P" in the amount of $500 and forged P's indorsement. X went to a Currency Exchange with a friend, Y. Y, as a favor, indorsed the check to aid X in cashing it. Y was well known at the Currency Exchange and had no knowledge that P's signature had been forged. The check was cashed but when Currency Exchange presented it to Drawee it was dishonored. Currency Exchange then neglected to notify Y of the dishonor, thereby discharging him on the instrument. Is Y liable to Currency Exchange for breach of any warranty under 3–417(2)? If Y indorsed as a favor and received no consideration, the answer is no. If, however, Y received all or part of the proceeds, either from Currency Exchange or X, the warranty of good title would be made and breached. Oak Park Currency Exchange, Inc. v. Maropoulos, 48 Ill.App.3d 437, 6 Ill.Dec. 525, 363 N.E.2d 54 (1977).

D. WARRANTY AGAINST MATERIAL ALTERATION

A person who buys a negotiable instrument usually assumes that he is entitled to recover according to the tenor of the instrument at the time of his purchase from the various parties whose signatures appear on the instrument at that time. If the instrument has been materially altered, however, the liability of the parties who signed before the alteration usually is limited to the tenor of the instrument before it was altered, and sometimes such liability is completely discharged. See 3–407. Primarily to protect transferees against losses that result in this way, each transferor who receives consideration warrants that "the instrument has not been materially altered." 3–417(2)(c). This warranty also helps protect against the risks that the transferee will be held liable for breach of the same warranty if he transfers the instrument or that he will be held liable for breach of the *presenters'* warranty against material alteration if he or a subsequent party obtains payment or acceptance.

> *Example 40*
> M makes a note for $9,000 payable to the order of P. P skilfully raises the amount to $29,000 and negotiates the note to A. A negotiates the note to HDC. The transferors' warranty against alteration was broken by P and A. HDC is entitled to collect $9,000 from M and the balance of the $29,000 from P or A; or HDC may rescind his contract with A and recover whatever consideration he paid A. Of course, P and A might be liable *secondarily* for the full $29,000.

E. WARRANTY REGARDING DEFENSES

This is the only warranty that differs depending on the nature of the transfer. Any transferor *other than a qualified indorser* warrants *absolutely* that "no defense of any party is good against him." 3–417(2)(d). If the indorsement is "without recourse," the qualified

Ch. 9   LIABILITY BASED ON WARRANTY

indorser's warranty is merely that he has *no actual knowledge* of any defense of any party good against him. 3-417(3), 1-201(25). To fill the gap left by the less than absolute liability of the qualified indorser, it is not unusual for the transferee who takes by a qualified indorsement to insist that the indorser give an express warranty regarding defenses generally or regarding some matter about which the transferee is particularly concerned. For example, the qualified indorser might be required to warrant absolutely that there are no defenses or that the maker or some prior indorser is not an infant.

Example 41

R, an adjudicated incompetent, draws a draft on E, payable to the order of P, on demand. By a qualified indorsement, P negotiates the draft to A. By an unqualified indorsement, A negotiates the draft to H. When H demands payment, R properly refuses on the ground of incompetency. Regardless of his knowledge of the defense, A, as an unqualified indorser, is liable to H for breach of the transferors' warranty against defenses. P, as a qualified indorser, is liable for breach of this warranty only if he had actual knowledge of the incompetency when he negotiated the note. If he broke the warranty, P is liable to H as well as A. If P had not indorsed, he would have been liable on the warranty regardless of his knowledge, but his warranty would have run only to his immediate transferee, A.

F.   WARRANTY OF NO KNOWLEDGE OF INSOLVENCY PROCEEDINGS

A transferor does not give any warranty that the party expected to pay or anyone else is a good credit risk or that he is solvent in the commercial sense. The transferee normally is expected to check these things for himself before he buys. However, a transferor for consideration does warrant that "he has *no knowledge* of any insolvency proceedings instituted with respect to the *maker or acceptor or the drawer of an unaccepted instrument.*" 3–417(2)(e). (Emphasis added) This warranty prevents a holder who is aware that such proceedings have been commenced from passing his almost certain loss to his transferee.

Example 42

R draws a draft on E, payable to the order of P. P negotiates the draft to A. Aware that P is insolvent, but unaware that insolvency proceedings have been instituted against P, A sells and delivers the draft to H. A broke no warranty. Even assuming that A had known that insolvency proceedings had been instituted against P, A would not have broken this warranty because P was the *payee*, not a maker, acceptor, or drawer.

G. HOLDER IN DUE COURSE AND TRANSFEROR'S WARRANTIES

The transferor's warranties extend to all transferees, including even a holder in due course in a case where breach of the warranty does not destroy any rights the holder in due course otherwise would enjoy. For example, a holder in due course of a stolen bearer instrument may rescind the transfer for breach of the warranty of title even though a holder in due course is not subject to a claim or defense based on nondelivery. Likewise, he can rescind for breach of the warranty

Ch. 9   LIABILITY BASED ON WARRANTY

against alteration, even though the alteration is the wrongful completion of an incomplete instrument and such completion does not impair his right to collect on the instrument as completed. Also, he can rescind for breach of the warranty against defenses although the defense is lack of consideration or some other defense which cannot be asserted against a holder in due course. Giving the holder in due course the advantage of these warranties protects him from being harassed by these or other defenses and from being required to prove his status as a holder in due course in a lawsuit not of his own choosing. Comment 9, 3–417.

§ 2. Presenters' Warranties

Anyone who pays or accepts an instrument is likely to do so on the basis of one or more assumptions. For example, a maker or acceptor who pays is likely to assume that the person obtaining payment is entitled to do so, that all of the signatures on the instrument are genuine, that the instrument has not been altered, that any documents delivered at the time of payment or acceptance are genuine and that he has no valid defense against the claim for payment. A drawee who pays or accepts is likely to make similar assumptions and to assume also that the draft is not an overdraft and does not otherwise exceed the bounds of his agreement with the drawer and that the drawer has not issued any stop payment order. An unqualified indorser or a drawer who pays assumes what other payors assume and, in addition, that the instrument has not been dishonored and that the other conditions to his own liability have been met. If a person pays or accepts because he is mistaken about any one of these or other matters, the law in some cases allows him to get his money back or rescind his acceptance, but in other cases it does not. This area of the law is very complicated, and courts, legislators and writers have had much difficulty in trying to develop and describe the governing legal principles.

The problem is not new. A leading case, and the starting point for most of the discussion of the law in this area, is Price v. Neal, 3 Burr. 1354, 97 Eng.Rep. 871, which came before Lord Mansfield in 1762. It involved two drafts. The drawee paid one after first accepting it, and he paid the other without having accepted it. Some time later he sought to recover both payments on the ground that the drawer's signature had been forged. Lord Mansfield decided that the drawee-acceptor could not recover either payment. In reaching this conclusion he said: "The plaintiff cannot recover the money unless it be against conscience in the defendant to retain it. * * * But it can never be thought unconscientious in the defendant to retain this money, when he has once received it upon a bill of exchange indorsed to him for a fair and valuable consideration, which he had bona fide paid, without the least privity or suspicion of any forgery."

For various reasons, American courts generally have followed and expanded the doctrine of Price v. Neal in similar or analogous cases involving mistaken payment or acceptance of commercial paper. For example, in Central Bank & Trust Co. v. G. F. C., 297 F. 2d 126 (5th Cir. 1961), a bank sued to recover a payment it had made to the payee of a check that had overdrawn the drawer's acount with the bank. The court denied the bank any right to recover the payment. Although the court did not impute negligence or other fault to the bank, it stated that in most cases of this kind the holder has no way of knowing of the paying party's mistake whereas the paying party can make such mistakes only if it is lax. The court indicated, however, that the basic reason for denying recovery in this and similar cases is that in the modern complicated business world, allowing recovery would often cause serious delay, uncertainty, and confusion, affecting not only the parties directly involved but many others as well.

A. THE DOCTRINE OF FINALITY

On the assumption that it is usually better to end the transaction on the instrument when it is paid or accepted rather than set aside a whole series of transactions when a mistake or wrongdoing is later discovered, the Code starts by adopting the broad principles established by Price v. Neal and progeny. It does so by providing that "payment or acceptance of any instrument is final in favor of *a holder in due course, or a person who has in good faith changed his position in reliance * * *"* (Emphasis added) 3–418 and Comment 1. Notice that the benefit of the doctrine of finality under the Code extends only to a holder in due course or one who has in good faith changed his position in reliance on payment or acceptance. A person who does not fall into one of these two categories is governed by general principles of restitution and so can almost always be required to give up any benefits received as the result of a material mistake. See Restatement of Restitution §§ 28–30 (1936). In short, only holders in due course and those who change their position in reliance on payment or acceptance get the benefit of the general rule that payment or acceptance is final even when made as the result of a material mistake. Even as to these favored parties, however, the general rule is subject to several exceptions. Aside from minor variations required by the special circumstances relating to bank deposits and collections, these exceptions are embodied in three implied warranties which relate to (1) the title of the presenting party, (2) the signature of the issuing party, and (3) material alteration. Normally, one who wishes to set aside a payment or an acceptance given to a holder in due course or to one who changed his position in good faith reliance must proceed on the basis of one of these three warranties which usually are given by *persons who obtain payment or acceptance and by prior transferors*. The "finality" doctrine in Section 3–418, therefore, is limited by any warranties made and breached upon presentment under Section 3–417(1).

Example 43

R draws a draft on E who is not indebted to him. P, payee, obtains the draft by fraud and negotiates the draft to H who is not a holder in due course because he knows the draft is overdue. Mistakenly believing that he (E) is indebted to R, E pays H. H deposits the proceeds in his savings account. When E discovers his mistake, he is entitled to recover the payment from H. Even though none of the presenters' warranties is broken, the doctrine of finality of payment does not apply because H is neither a holder in due course nor a good faith relier who changed his position as required by Section 3–418. According to ordinary principles of restitution, E is entitled to recover the payment which he made in good faith as the result of a material mistake. Since none of the presenters' warranties was broken, if H had been either a holder in due course or one who changed his position in good faith reliance on the payment, the doctrine of finality embodied in Section 3–418 would have applied and E would not have been entitled to recover the payment.

B. WARRANTY OF TITLE

Warranties on presentment are given by "any person who obtains payment or acceptance" of an instrument and "any prior transferor" to a "person who in good faith pays or accepts." 3–417(1). The first warranty is that the presenter "has good title to the instrument or is authorized to obtain payment or acceptance on behalf of one who has a good title." 3–417(1)(a). Note that no warranty of "rightful transfer" is made when an instrument is paid and accepted. Compare 3–417(2)(a).

The "title" warranty protects one who pays or accepts for the wrong party, usually someone who is not a holder. If payment has been made, the remedy for breach is to recover the amount paid, normally the face amount of the instrument.

Example 44

M issues a note payable to P's order in the amount of $1,000. T steals the note, forges P's indorsement, and sells the note to A. Thinking A is the holder, M pays A. (A is not the holder because the note runs to P.) Assuming that P is not precluded from asserting the unauthorized signature, M, having paid on the forged indorsement, is liable to P for conversion, 3–419(1)(c). The conversion loss is presumed to be the face amount of the instrument. 3–419(2). M is entitled to recoup his loss from A on the basis of A's breach of the presenters' warranty of title. A had neither title nor proper authority to obtain payment. If A is unavailable or without funds, M may recoup his loss from T on the basis of T's breach of the presenters' warranty of title, which is given by prior transferors. See 3–417(2)(a).

This warranty may also be helpful to a drawee who is induced to accept a draft presented by someone who lacks title. The warranty allows him to rescind his acceptance, thus avoiding liability to the presenting party and simultaneously avoiding the risk of becoming liable to some later party who purchases the instrument in reliance on the acceptance. If the instrument does fall into the hands of a holder in due course before the acceptor can rescind his acceptance, he will be required to pay; but, in this case, the warranty of title gives him the right to recover his loss by suing the party who obtained the acceptance.

Arguably, the warranty of title may be broken even though the party who obtains payment or acceptance is in fact the *holder* of the instrument. For example, although a finder or a thief of a bearer instrument has possession and is a holder, he nonetheless lacks title. At least one court, however, has held that the presenter's title warranty is limited to the validity of signatures in the chain of necessary indorsements. Sun 'N Sand, Inc. v. United California Bank, 21 Cal.3d 671,

148 Cal.Rptr, 329, 582 P.2d 920 (1978). This warranty normally is not important in these circumstances because a person usually can discharge his liability on an instrument by paying the holder whether or not he has title. See 3–603(1).

It is interesting to consider this warranty in the light of the doctrine of Price v. Neal. Under that doctrine, payment to or acceptance for one acting in good faith is treated as equal to an admission of the genuineness of the *drawer's* signature; consequently, the drawee may not thereafter complain that the *drawer's* signature has been forged. In contrast, the warranty of title enables the paying or accepting party to relief whenever the presenting party lacks title because an indorsement of the *payee or some other party in the chain of title* has been forged. This difference in treatment is said to be justified by the fact that normally a drawee can avoid loss when the *drawer's* signature is forged simply by comparing the signature on the instrument with one it has on hand, whereas a party making payment or acceptance normally has no reasonable opportunity to make a similar comparison to determine the genuineness of an *indorser's* signature. Comment 3, 3–417.

Many scholars doubt the validity of this distinction especially in the modern business world where there is really no practical way of checking all signatures of makers or drawers consistently with the need for speed. At present, a sounder reason might be that it tends to let the loss remain where it first falls among innocent parties rather than upsetting a series of transactions with all of the waste of time and effort

C. WARRANTY AGAINST KNOWLEDGE SIGNATURE OF MAKER OR DRAWER IS UNAUTHORIZED

Those who obtain payment or acceptance and prior transferors do not warrant *absolutely* that the signature of the person named as the issuing party is genuine or authorized. At most, they warrant that they have *no knowledge* that the "signature of the maker or drawer is unauthorized." 3–417(1)(b). "Knowledge," for this purpose means "actual" knowledge. 1–201(25).

To understand the special nature and effect of this second presenters' warranty it should be noted that it relates only to the signature of the *maker* or *drawer* —the parties who normally *issue* an instrument. The only other signature that might concern a person who is paying or accepting is that of an indorser in the chain of title, but the authenticity of the signature of an indorser in the chain of title is already covered by the warranty of title discussed above.

Example 45

M makes a 60-day note and delivers it to P, the payee, for goods sold. A week later, P offers to sell the note to A but A says he will buy the note only if N first indorses it. P asks N to indorse. N refuses. P forges N's indorsement. P then indorses the note and delivers it to A for value. A negotiates the note to B for value. At maturity, B obtains payment from M. At the time B knows that N's signature has been forged. B has broken no warranty. N is neither a maker nor drawer so the warranty relating to their signatures is not broken. N is not in the chain of title so the warranty of title is not broken. Also, M suffers no damage because he was discharged by his payment in good faith to B, the holder.

By providing by implication that the warranty is not broken by one who obtains payment from a drawee or acceptor *without knowledge* that the drawer's signature has been forged, the Code in effect adopts the holdings regarding both of the drafts in the case of Price v. Neal. The Code goes beyond these holdings, however, by implying also that one who obtains payment from a maker, drawer, or indorser or acceptance from a drawee without knowledge that the signature of the maker or drawer has been forged does not break this warranty.

In contrast, when a person obtains payment or acceptance *with knowledge* that the signature of the maker or drawer has been forged, this warranty usually is broken. There are two exceptions, and both are limited to cases where a holder in due course obtains *payment* while acting in good faith.

The first exception is that this warranty is not given or broken when a holder in due course obtains payment from a *maker or drawer in good faith while knowing that the paying party's signature is forged.* 3–417(1)(b)(i). It is reasoned that a maker or drawer should know his own signature, and if not, he should bear whatever loss or inconvenience results from his mistake, rather than thrust it on a holder in due course who has obtained payment in good faith. However, this exception is recognized regardless of whether the payor is negligent.

One must not assume that this exception is broader than it actually is. First, a person cannot be a holder in due course if he knows when he acquires an instrument that the signature of the maker or drawer is unauthorized. Also, even assuming that a person acquires an instrument as a holder in due course, he does not act in good faith if he knows when he obtains payment that the payor is not legally obligated to make it. Knowledge that the signature of the maker or drawer is unauthorized does not negate his good faith, however, if the holder in due course has good reason to believe that the maker or drawer is legally obligated to pay him, as would be true if the holder in due course thinks that the negligence of the maker or drawer substantially contributed to the unauthorized signature. (See 3–406).

The second exception is made if the holder in due course takes a draft after it has been accepted *or* obtains acceptance without knowledge that the drawer's signature has been forged, but later learns of the forgery and obtains payment from the acceptor in good faith. 3–417(1)(b)(iii). In this situation, it is sufficient to establish the good faith of the holder that he has relied on the acceptance because in this case he rightly feels that the acceptor should pay. If the acceptor had refused to accept when it was presented by a prior party, the holder in due course probably would not have purchased the instrument. If the acceptor had refused to accept when the draft was presented by the holder in due course, the latter would have been on notice of the unauthorized signature sooner and would have been in a better position to pursue his transferor or some prior party. This second exception goes further than the holding of Price v. Neal, in which case it appears, not only that the defendant was a holder in due course by our modern standards, but also that he obtained payment *without knowledge* of the forgeries.

Example 46

P forges R's signature on a draft and negotiates it to HDC. Before maturity, HDC, unaware of the forgery, obtains E's acceptance. HDC then learns of the forgery. Nonetheless, he honestly believes that he is entitled to be paid and obtains payment from E. E is not entitled to recover the payment from HDC. A holder in due course who obtains an *acceptance* in good faith, being unaware that the drawer's signature is unauthorized, does not give or break the warranty of no knowledge if later, having learned of the forgery, acting in food faith, he obtains *payment* from the acceptor.

In contrast, whenever a holder in due course *obtains an acceptance* with actual knowledge that the drawer's signature has been forged, he gives and breaks the warranty. His good faith, if it exists, does not help him. Furthermore, if he later obtains payment from the acceptor on the basis of such an acceptance, he again breaks this warranty.

Example 47

P forges R's signature to a draft. P negotiates the draft to HDC. HDC then learns of the forgery. Nonetheless, HDC obtains E's acceptance. At maturity, HDC obtains payment from E. Even assuming that HDC obtained the payment in good faith, E is entitled to recover it. HDC broke the warranty that he had no knowledge that R's signature was unauthorized when he obtained the acceptance, and later when he obtained payment. Also, HDC would have broken the warranty if, without having obtained acceptance, he obtained payment from E knowing that R's signature had been forged.

D. WARRANTY AGAINST MATERIAL ALTERATIONS

Finally, a party who presents an instrument for payment or acceptance and any prior transferor give an *absolute* warranty—that is, one which *does not depend upon knowledge* —"that the instrument has not been materially altered." 3–417(1)(c). Again there are two exceptions, both of which parallel the exceptions relating to the second warranty and apply only in cases where a *holder in due course obtains payment in good faith.*

The first exception is that a holder in due course who obtains payment in good faith does not give this warranty to a *maker* or *drawer.* 3–417(1)(c)(i and ii). A maker or drawer who pays an altered instrument is as likely to be at fault as when paying an instrument on which his signature is unauthorized; but, once again, the exception applies regardless of fault.

Of course, good faith again poses a problem. On the one extreme, if the holder in due course himself made the alteration, his bad faith in obtaining payment is clear and so he gives and breaks the warranty. On the opposite extreme, if a holder in due course is total-

Ch. 9 LIABILITY BASED ON WARRANTY

ly unaware of the alteration when he obtains payment, his good faith is clear and so he would neither give nor break the warranty. Knowledge of the alteration at the time of obtaining payment might or might not negate good faith. Normally it does. It does not, however, if the holder in due course has good reason to think that the maker or drawer is legally or morally bound to make the payment in question. For example, since parties to an instrument remain liable to a holder in due course according to the original tenor of an instrument which has been altered by deletion, addition, or change, he does not show lack of good faith or give the warranty when he obtains this limited payment while aware of the alteration. Similarly, he does not show lack of good faith or give the warranty despite knowledge of the alteration if he obtains payment from the maker or drawer according to the terms of the altered instrument, if he has reason to think that the maker's or drawer's negligence has substantially contributed to the alteration or if he knows that the alteration consists of wrongfully completing an incomplete instrument. In such cases, a holder in due course has a right to payment according to the terms of the altered instrument. See 3–407(3).

The second exception is recognized when a holder in due course, acting in good faith, obtains payment from an *acceptor* who had accepted the instrument *before* it was acquired by the holder in due course. This exception is recognized whether the alteration occurred before or after the acceptance. 3–417(1)(c)(iii), (iv). If the alteration occurred before the acceptance, it is recognized even though the acceptance expressly states that the instrument will be paid as originally drawn.

It seems clear that a holder in due course who purchases an instrument which already has been accepted acts in good faith and so does not give this warranty if he obtains payment from the acceptor while unaware of the alteration. Even though he is aware of the alteration when obtaining payment, he may act in good faith if he has good reason to think that the alteration was made *before* the instrument was accepted. In this case, he buys the instrument relying on the assumption that the acceptor is liable according to its terms as altered, and the acceptor is in part responsible for this assumption. A holder in due course does not act in good faith, however, and hence he gives and breaks this warranty if he obtains full payment according to the altered terms knowing that the alteration was made *after* the acceptance. In this case he has no basis for assuming that the acceptor is liable except according to the terms of the instrument at the time of his acceptance.

If the holder in due course, himself, *obtains the acceptance*, he gives the warranty just like anyone else, and the acceptor can hold him liable if the instrument has been altered before it is presented. This is so whether or not the holder in due course knows of the alteration when he obtains acceptance. Similarly if the holder in due course, having obtained acceptance of an altered instrument, later obtains payment, he gives and breaks this warranty just like anyone else.

Ch. 9 *LIABILITY BASED ON WARRANTY*

It should be emphasized that normally the only warranties given by a person who obtains payment or acceptance relate to title, the signature of the issuing party, and alterations. For example, no warranties are given to protect a drawee who pays an overdraft or who pays in disregard of a stop payment order. Of course, the presenting party might give an *express* warranty with respect to any matter normally not covered.

E. PRESENTERS' WARRANTIES GIVEN ALSO BY PRIOR TRANSFERORS

As mentioned earlier, the three presenters' warranties that have been discussed are given not only by those who obtain acceptance or payment but also by prior transferors. 3–417(1).

Example 48

M issues a note to P. P fraudulently raises the amount and negotiates the note to A. A negotiates the note to HDC. At maturity, HDC acting in good faith obtains payment from M. Since HDC obtained payment from the maker in good faith, he did not make or break the presenters' warranty against material alteration. 3–417(1)(c)(i). Although A is a prior transferor, M cannot recover from A for breach of A's *transferors'* warranty against material alteration. The reason is that transferors' warranties run only to the transferor's immediate *transferee* and any subsequent *holder* who takes the instrument in good faith; and M is neither. As a prior transferor, however, A is liable to M on the basis of the *presenters'* warranty against material alteration unless he, A, is also a holder in due course who acted in good faith.

3–417(1)(c)(i). If M cannot recover from either HDC or A, M can recover for breach of the presenters' warranty against alteration from P, a prior transferor who is clearly not a holder in due course who acted in good faith.

F. TRANSFEREES OF HOLDER IN DUE COURSE

The special advantage given holders in due course concerning warranties relating to the signature of the issuing party and alterations extends also to their transferees under the "shelter" provision of the Code. 3–201(1).

Example 49

M issues a note to P. P alters the note to increase the sum payable and then negotiates it to HDC. HDC negotiates the note to T who is not a holder in due course in his own right because he knows the note is overdue. T, acting in good faith and unaware of the alteration, obtains payment from M. T has not given or broken the warranty against material alteration because he has the *rights* of a holder in due course who does not give this warranty when he obtains payment from a maker while acting in good faith.

§ 3. Remedies Available for Breach of Warranty

There are a number of generalizations that can be made without distinguishing between warranties made by transferors and those made by presenters.

To begin, there can be no relief on the basis of a breach of either type of warranty without good faith reliance. For example, if a person purchases an instrument knowing that the amount has been raised, he cannot obtain relief on the basis of the *transferors'* warranty against alteration and a person who pays an instrument knowing that it has been altered cannot obtain relief on the basis of the *presenters'* warranty against alteration. Also, breach of either class of warranty normally gives the injured party the right to elect between rescinding the transaction and taking back whatever he parted with or letting the transaction stand and suing for damages.

Regardless of whether a transferors' or a presenters' warranty is broken, if the aggrieved party elects to recover damages he is limited to actual damages that are reasonably foreseeable by the party who broke the warranty. The general object is to put the aggrieved party in the position that he would have been in monetarily if there had been no breach. For example, if a transferor's warranty is broken by a transferor without indorsement (who incurs no secondary liability) the transferee can usually recover the difference between the value of the instrument as it was and as it would have been if there had been no breach. If the signature of the maker, acceptor, or drawer, or an unqualified indorser in the line of title, is forged, this may be the face amount of the instrument. But if no one else is liable on the instrument and the person whose name has been forged was insolvent when payment was due, the amount of recovery normally would be limited to what might have been recovered from that person if he had actually signed. Also, any incidental or consequential damages that are reasonably foreseeable may be recovered. This would include the reasonable costs of litigation to which the aggrieved party is necessarily subjected as the result of the breach of warranty. Regardless of which type of warranty is involved, if a person is acting as a selling broker, the extent of his liability for breach of warranty depends on whether he discloses this fact. If he does, he warrants only his good faith and authority; if he does not, he is liable on either a transferors' or presenters' warranty just as if he were acting as principal. 3–417(4). As is true of warranties generally, either transferors' or presenters' warranties can be eliminated by express agreement. If a transfer is made by indorsement, however, a disclaimer of liability must be made in the indorsement; otherwise proof of the disclaimer would violate the parol evidence rule. Finally, by indorsement or otherwise a transferor or presenter may add one or more express warranties. For example, a transferor may warrant that a primary party is solvent.

§ 4. Warranties in Check Collection Process

Section 4–207 of Article 4 regulates the warranties made by customers and collecting banks in the check collection process. The warranties are similar to those made under Section 3–417. Thus, customers and collecting banks who obtain "payment or acceptance" of an item make "presenters" warranties under Section 4–207(1) and those who transfer an item for a "settlement or other consideration" make warranties of transfer under Section 4–207(2). There are important differences, however, and Section 4–207 is, in addition, meshed with the complicated provisions of Article 4.

Accordingly, we will postpone the discussion of Section 4–207 to Chapter 13.

CHAPTER 10

HOLDERS IN DUE COURSE

§ 1. Importance

Although the elements essential to liability are present, a defendant may possibly avoid liability by proving one or more defenses. The range of defenses which might be asserted effectively depends largely on whether or not the plaintiff has the rights of a holder in due course. If he has, he holds the instrument free of most defenses of prior parties. 3–305(2). He also holds the instrument free of any prior party's right to claim the instrument itself even if it was obtained by fraud, theft, or other illegal means from that person. 3–305(1). He enjoys freedom from liability based on warranties under some circumstances in which others would be held liable. See 3–417(1)(b) and (c). He has the advantage of being able to transfer his special rights and privileges to others. 3–201(1). These and other advantages are mentioned in various scattered sections of the Code. Earlier the requirements of a holder in due course were described in broad terms. Because of their vital importance in many cases, they will now be described more fully.

§ 2. Requirements

The basic requirements are stated in Section 3–302(1) of the Code which provides:

> A holder in due course is a holder who takes the instrument
>
> (a) for value; and
> (b) in good faith; and
> (c) without notice that it is overdue or has been dishonored or of any defense against or claim to it on the part of any person.

Section 3–302(1) should be considered carefully and remembered because it is the logical starting point when necessary to determine whether a party is a holder in due course. In many cases, however, it is necessary to proceed further to find the answer. (For a discussion of a few special situations in which a holder is not a holder in due course even though he satisfies the requirements of Section 3–302(1) set forth above, see pages 267–268.)

§ 3. Instrument

The "instrument" referred to Section 3–302(1) is a "negotiable instrument." 3–102(1)(e). Although, as was explained earlier (see page 10), an instrument that fails to be negotiable only because it is not payable to bearer or to order is governed by Article 3 in other respects, *there can be no holder in due course of such an instrument or of any other instrument that is not a negotiable instrument as defined by Section 3–104(1).* 3–805.

§ 4. Holder

An easily forgotten truism is that *one cannot be a holder in due course unless one is a holder.* As explained earlier, a *holder* is a person who is in possession of an instrument that *runs* to him; and a bearer instrument runs to whoever is in possession of it, whereas an order instrument runs to the payee if it has never been indorsed, and to the last indorsee if it has. (See pages 103–104.

§ 5. Value

The Code contains two overlapping but clearly different definitions of value. A general definition, contained in Article 1, defines value broadly as including, among other things, "any consideration sufficient to support a simple contract." 1–201(44). Article 3, which governs commercial paper, defines value much more narrowly. This definition will be explained shortly. Here it is sufficient to recognize that *throughout our discussion of commercial paper, value is used only in the narrow sense of* Section 3–303.

In the law of commercial paper, *consideration* has the same meaning that it has in the law of simple contracts. Its meaning is quite different from the meaning of value in the sense in which it is used in the law of commercial paper. Equally important, in the law of commercial paper each of these terms has a special legal significance of its own.

The primary legal significance of value is that unless a holder has taken an instrument for value, he

Ch. 10 *HOLDERS IN DUE COURSE*

cannot be holder in due course. 3–302(1)(a). Value is therefore an essential element in establishing this highly desirable *status*. The primary legal significance of consideration is that lack or failure of consideration is a *defense* which can be asserted by a defendant who is sued on an ordinary contract, on an instrument that is not negotiable, or on an instrument that is negotiable, by a plaintiff who lacks the rights of a holder in due course. 3–306(c), 3–408. Both concepts have less important legal significance in other areas of commercial paper. For example, unless the plaintiff or some prior party has given *value* before an instrument is overdue, there is normally no right to recover from an accommodation party 3–415(2); and the transferors' warranties are given only by a transferor who receives *consideration*. 3–417(2). The meaning of each term, however, remains the same in the various contexts in which it is used.

The starting point for determining the meaning of value in the narrow sense in which it is used in the law of commercial paper is Section 3–303 which provides:

A holder takes the instrument for value

(a) to the extent that the agreed *consideration has been performed* or that he acquires a security interest in or a lien on the instrument otherwise than by legal process; or

(b) when he takes the instrument in payment of or as security for an antecedent claim against any person whether or not the claim is due; or

(c) when he gives a negotiable instrument for it or makes an irrevocable commitment to a third person. (Emphasis added.)

[*240*]

An examination of the foregoing provision of the Code shows that although the legal significance of value and consideration are distinctly different, to a certain extent the key to the meaning of value is consideration.

Since the Code does not define consideration, it usually will be necessary for the courts to refer to cases decided in their own jurisdictions to determine its meaning and, consequently, the meaning of value. Consideration has been defined by the courts and writers in various ways. A reasonably workable definition is that "consideration consists of doing or *promising* to do what one is not already legally bound to do, or refraining or *promising* to refrain from doing what one has a legal right to do, in return for a promise." A somewhat shorter definition might be "giving up or *promising* to give up a right, power, or privilege, in return for a promise." The key ingredient is that the "consideration" must be bargained for and given in exchange for the promise. Restatement, Second, Contracts § 71.

A. EXECUTORY PROMISE NORMALLY NOT VALUE

Both foregoing definitions recognize that *consideration* might consist of a *promise*, even though the promise remains executory, that is, unperformed. Referring to Section 3–303, and particularly the phrase "to the extent that the agreed consideration *has been performed*," it will be seen that normally an executory promise cannot constitute *value*.

Ch. 10   *HOLDERS IN DUE COURSE*

The foregoing general principles relating to value and consideration are illustrated by the following case:

Example 50

M issues a note for $1,000 to P as a gift. Since P gives up nothing in return for M's promise in the note, M receives no consideration and can assert this as a defense if sued by P or any later party who is not a holder in due course. Next, assume that P negotiates the note to X in return for X's promise to pay $900. Since X was not previously legally bound to pay $900, his promise to do so is consideration for the obligation of P as transferor. However, as long as X's promise is unperformed, X has not given value and so is not a holder in due course. Consequently, if X sues M, he can be defeated by M's defense, lack of consideration. Assume that X negotiates the note to Z who pays $900 cash. Since his consideration has been completely executed, Z has given value, and if he meets the other requirements, he is a holder in due course. As such, he is entitled to recover from M at maturity because M's defense—lack of consideration—is not good against a holder in due course. Notice that Z can recover the face amount of the note, $1,000, even though he paid only $900 for it.

B. NOTICE OF CLAIM OR DEFENSE WHILE PROMISE IS EXECUTORY

If an instrument is negotiated to a party in return for his promise, that party does not become a holder for value immediately. However, by fulfilling his promise, he can become a holder for value and consequently a holder in due course. If he learns of a defense or claim before he has fully carried out his promise, he can become a holder in due course only to the extent that his promise has been carried out when he learns of the defense or claim. 3–303(a). For example,

Example 51

P induces M by fraud to deliver a note for $1000. Before maturity P negotiates the note to H, who acts in good faith, in return for H's promise to pay $1000 in four equal monthly instalments. After paying three instalments, H learns of P's fraud. If H nonetheless pays P the remaining $250, he will be entitled to recover at most only $750 from M at maturity because H can be a holder in due course only to that extent. If H had agreed to pay only $750 as the total consideration for the instrument in three equal instalments and paid the $750 before learning of P's fraud, H might have become a holder in due course to the fullest possible extent and as such entitled to recover the full $1000 from M.

C. EXECUTORY PROMISE AS VALUE

The basic reason for not recognizing a holder as a holder in due course, except to the extent that his promise has been performed before he learns of a defense or claim, is that to the extent that his promise is executory a holder who learns of a defense or claim normally can avoid his loss by simply refusing to carry out his promise. If he is unable to avoid loss in this way, it becomes necessary to accord him the status of a holder in due course to protect him although his promise given in return for the instrument is still executory. The Code expressly recognizes this in two situations. 3–303(c).

Ch. 10 *HOLDERS IN DUE COURSE*

The first situation arises when a holder acquires an instrument in return for an executory promise that constitutes an irrevocable commitment by the holder to a third party. For example, suppose that a bank credits a depositor's account with the amount of a check and then, while unaware of a defense on the deposited check, certifies a check drawn by the depositor for the balance of the account. In this case the bank is held to have given value for the deposited check even though it does not pay the certified check until after it has learned of a defense on the deposited check. See Freeport Bank v. Viemeister, 227 App.Div. 457, 238 N.Y.Supp. 169 (1929).

The second situation occurs if a holder issues his own negotiable instrument in return for the instrument on which he seeks payment; that is, there is an exchange of negotiable instruments. For example, suppose a bank gives its own draft in return for the note of a third party, and after learning of a defense on the note, it pays the draft to a transferee who is a holder in due course. In this case justice obviously requires that the bank should not be denied the status of holder in due course merely because it learned of a defense before making a payment it was bound to make. See First Nat'l Bank of Waukesha v. Motors Acceptance Corp., 15 Wis.2d 44, 112 N.W.2d 381 (1961). The Code goes beyond this, however, and provides that a person becomes a holder for value of an instrument for which he issues his own negotiable instrument even though he is never called upon to pay a holder in due course. It is reasoned that the possibility of the instrument falling into the hands of a holder in due course furnishes a sufficient basis for giving him this protection. Comment 6, 3–303.

D. BANK CREDIT AS VALUE

In the typical case in which a bank receives one or more negotiable instruments for deposit in a customer's account, the bank immediately credits the customer's account, but the depositor acquires no right to draw against the items and the bank need not permit him to do so until the item is collected. As might be expected, the bank does not become a holder for value merely by crediting the depositor's account. If the depositor is permitted to draw against a deposited item before it is collected, however, the bank immediately becomes a holder for value concerning that item. 4–208(1)(a). Consequently, if it becomes necessary for the bank to sue on the item, the bank normally enjoys the status of holder in due course unless it had learned of a defense or claim before allowing the withdrawal. This is true even though the deposit is made under a restrictive indorsement such as "for deposit only" and the deposit slip expressly states that the bank is acting only as agent for collection. It is reasoned that the provision is for the protection of the bank and that the bank may waive the protection. First National Bank of Somerset County v. Margulies, 35 Misc.2d 332, 232 N.Y.S.2d 274 (1962).

When a bank credits a depositor with a series of deposits and permits a number of withdrawals over a period of time, it is not always easy to determine whether the depositor has been permitted to withdraw funds credited on the basis of a particular item before the bank learns of a claim or defense on the item. In tracing the deposits to withdrawals, the Code applies the rule of "first in, first out" or "fifo" as it is sometimes called. 4–208(2), 4–209. For example:

Example 52

> D's account in X Bank starts with a zero balance. On three successive days, D deposits in his account checks for $100 each by A, B, and C in that order. X Bank immediately credits D's account as each check is deposited. After D withdraws $200, X Bank learns that the checks of B and C are subject to defenses. If X Bank subsequently permits D to withdraw the remaining $100, it will nonetheless be treated as having given value for B's check before receiving notice because the deposit based on this check was withdrawn when D obtained the $200. However, since the deposit of C's check occurred last, it is presumed to have been withdrawn last. Consequently, the bank is deemed not to have given value for C's check until after it received notice of C's defense. Therefore, in an action against C the bank could not enjoy the status of a holder in due course.

In contrast with the typical case involving the deposit of one or more negotiable instruments as described above, a bank may acquire a negotiable instrument under circumstances that give the depositor or transferor a *right* to draw against it and impose on the bank a *duty* to allow this. Whether this right and the corresponding duty arise from an express contract between the parties or from rules governing the collection process, the bank is considered as having given value immediately when the right arises and before any withdrawal against the item has been made. 4–208(1)(b). Thus another exception occurs to the general rule that an executory promise does not constitute value.

E. TAKING INSTRUMENT AS PAYMENT OR SECURITY FOR ANTECEDENT DEBT AS VALUE

An executory promise is the only form of *consideration* which normally does not constitute *value*. Conversely, the Code recognizes at least one situation in which a holder who has not furnished consideration in the usual sense still is treated as having given value. This exception occurs if a debtor, not legally bound to do so, negotiates to his creditor *as security* an instrument issued by a third party and does not require the creditor to extend the time for payment of the debt or to give up anything else in return for the instrument given as security.

This exception is provided by Section 3–303(b), which states that a holder takes an instrument for value "when he takes the instrument *in payment* of or *as security* for an antecedent claim against any person whether or not the claim is due." (Emphasis added.) If the instrument is taken in *payment* for a debt there clearly is consideration because the creditor gives up his right to the original debt. If he takes it merely as *security* and gives up no rights, he furnishes no consideration. He is treated as giving *value*, however, and so may qualify as a holder in due course. 3–303(b). If there is no antecedent debt to be satisfied or secured, however, the holder does not give value.

Ch. 10   *HOLDERS IN DUE COURSE*

See Quazzo v. Quazzo, 136 Vt. 107, 386 A.2d 638 (1978).

Example 53

> P holds a draft accepted by A. P negotiates the draft to X, as security for a debt owed by P to X. X does not agree to extend the time for payment or incur any other detriment in return for the draft. When P fails to pay the secured debt, X sues A as acceptor. A asserts a defense that is good only against one who is not a holder in due course. Judgment for X. Although X did not incur a detriment and so did not furnish *consideration* for the draft, he gave *value* by taking the draft as security for an antecedent debt. Assuming that X met the other requirements, X is a holder in due course and not subject to A's defense.

A person acquiring a lien against a negotiable instrument by virtue of *legal process* and not by voluntary action of the owner of the instrument is not considered to have given value. 3–303(a).

Suppose a creditor receives a check of a third party from his debtor with the understanding that he will attempt to collect the proceeds and will apply the funds, if any, upon the transferor's debt. In Wilson Supply Co. v. West Artesia, 505 S.W.2d 312 (Tex.Civ.App. (1974)) it was held that the creditor did not pay value because the check was not received *in payment*. The dissent argued that the check was taken both in payment of and as security for an antecedent claim under Section 3–303(b).

§ 6. Good Faith

Since at least as early as Miller v. Race, 1 Burr. 452, 97 Eng.Rep. 398 (1758), the courts, and more recently, the legislatures, have been trying to develop a definition of good faith that is satisfactory to all concerned. That they have not been wholly successful is attributable mainly to disagreement over the extent to which the test of *good faith* should be subjective, depending only on what is found to be the state of the holder's mind, or *objective*, depending on other facts. As seen above, the requirement that a holder give *value* is totally objective, having nothing to do with the holder's state of mind when he takes the instrument. As will be seen, the requirement of taking *without notice* sometimes depends on the holder's state of mind and sometimes on extrinsic circumstances; thus it is both subjective and objective.

The 1952 Edition of the Code required "good faith," "including observance of reasonable commercial standards of any business in which the holder may be engaged." This latter requirement was deleted from later editions because it seemed to imply an objective rather than a subjective test of good faith. The test of good faith as it is now stated in the Code is entirely subjective. The Code provides that good faith means "honesty in fact in the conduct or transaction concerned." 1–201(19). Taken literally, this does not require either diligence or prudence. See Riley v. First State Bank, 469 S.W.2d 812 (Tex.Civ.App.1971) A gullible person might be found to have acted in good faith in circumstances where a shrewd person would not.

It protects the person with the "pure" heart and the "empty" head.

Under the prevailing subjective test, a holder is not necessarily barred from being a holder in due course merely because he receives the instrument from a total stranger or for substantially less than its face value, or because in some way he has failed to observe reasonable standards of the business community. Nor is the requisite good faith negated by the fact that *after* he takes the instrument in circumstances establishing good faith and the other requisites of being a holder in due course, he learns facts that make him suspicious. Manufacturers & Traders Trust Co. v. Murphy, 369 F.Supp. 11 (W.D.Pa.1974).

It is said that the subjective test promotes the policy of encouraging the free transferability of commercial paper. In some cases, however, it has appeared to run counter to the general policy of protecting consumers from unfair or deceptive practices; and some courts and legislatures have reacted against it. The operative test in commercial cases, at least, is whether what the holder in fact knew imposed a duty of further inquiry. The answer is no unless the failure to inquire indicates a deliberate desire to evade knowledge because of a belief that further investigation would disclose an actual claim or defense. Given the asserted interest in the "free flow of commercial paper," these circumstances must be compelling indeed. See Chemical Bank of Rochester v. Haskell, 51 N.Y.2d 85, 432 N.Y.S.2d 478, 411 N.E.2d 1339 (1980) (given actual knowledge, would a "commercially honest" person take without further inquiry?)

Example 54

P entered a written contract with M to install cooling equipment on M's premises. M was to pay when the work was done. M issued to P a negotiable note, promising to pay the contract price at a fixed date. P orally agreed not to discount or negotiate the note to a third party but this restriction did not appear on the instrument. Shortly after the work began, M learned that P intended to negotiate the note to Bank. M telephoned Bank and informed them of the agreement prohibiting negotiation. Nevertheless, P negotiated the instrument to Bank for value. Thereafter, P went bankrupt and failed to complete performance under the contract. Bank sued M on the note and M raised the defense of failure of consideration. Held, Bank was a holder in due course and took the instrument free of the defense. The defense arose after Bank took the instrument. The knowledge of the restriction communicated before the instrument was transferred was, without more, insufficient to require any further investigation. The note was negotiable on its face and there was no evidence that Bank knew of P's financial condition or had any information about the probability that P would or would not complete the contract. On the other hand, if Bank knew that P was insolvent or that P had failed to complete other contracts, the failure to investigate further might indicate a deliberate desire to evade further knowledge. See Factors & Note Buyers, Inc. v. Green Lane, Inc., 102 N.J.Super. 43, 245 A.2d 223 (1968).

§ 7. Without Notice

Even though a holder acquires an instrument for value and in actual good faith, he does not qualify as a holder in due course unless he also takes the instrument "without notice" of any of three things: (1) that it is overdue, (2) that it has been dishonored, or (3) that there is a defense against it or a claim to it by any person. 3–302(1)(c).

Any one of these three things should serve as a danger signal to a person who is contemplating taking an instrument for value. It should warn him that he may be buying a lawsuit and that he should not expect to occupy the favored position of holder in due course.

Of course, a person should be barred from being a holder in due course by *actual* notice of any of these things. If he sees the danger signal but chooses to disregard it, he must take the consequences.

But actual notice is not always necessary to bar a person from being a holder in due course. This is clearly established by Section 1–201(25) which provides that a person has "notice" of a fact when "(a) he has *actual knowledge* of it; or (b) he has *received* a notice or notification of it; or (c) from all the facts or circumstances known to him at the time in question he *has reason to know* that it exists." (Emphasis added) It is also established by Section 1–201(26) which provides that "a person '*receives*' a notice or notification when (a) it comes to his attention; or (b) it is duly delivered * * * at any * * * place held out by him as the place for receipt of such communications." (Emphasis added) Section 3–304 of the 1978 edition of the Code amplifies the "without notice" condition of Section 3–302(1)(c). The first three subsections state when a purchaser has "notice" and subsection 4 states when a purchaser does not have notice even though he has knowledge of certain facts. New York, however, has added a subsection 7 to Section 3–304, which provides that the purchaser must have "knowledge of the claim or defense or knowledge of such facts that his action in taking the instrument amounts to bad faith." This variation was enacted "in order to make it clear that the subjective test, applicable in this state even before the enactment of the Code was intended to be continued." Chemical Bank of Rochester v. Haskell, 51 N.Y. 2d 85, 432 N.Y.S.2d 478, 411 N.E.2d 1339 (1980).

To bar a purchaser from becoming a holder in due course, however, notice must be received at such time and in such manner as to give a reasonable opportunity to act on it. 3–304(6). If notice is given to an organization, it is effective when it is brought to the attention of the person conducting the transaction or when it would have been brought to his attention with the exercise of due diligence, whichever is sooner. 1–201(27). For example, a notice received by a bank president a minute before a teller cashes a check is not likely to be effective to bar it from being a holder in due course. Comment 12, 3–304. But if the president acts swiftly and succeeds in bringing the notice to the teller's attention before he cashes the check, the bank is charged with notice and cannot be a holder in due course of the check.

Although a public filing or recording is often binding notice of a prior claim to a property interest so as to bar a later purchaser or lienor of goods or real property, such recording is not effective to charge a purchaser of a negotiable instrument with notice so as to bar him from becoming a holder in due course.

Suppose that notice is received but is forgotten before the instrument is purchased. Can the purchaser be a holder in due course? Prior to the Code, courts were divided. See Graham v. White-Phillips Co., 296 U.S. 27, 56 S.Ct. 21, 80 L.Ed. 20 (1935). The Code avoids taking sides by providing that "the time and circumstances under which a notice or notification may cease to be effective are not determined by this Act." 1–201(25).

Because the requirement of good faith is so often confused with the requirement of taking without notice, perhaps it should be emphasized that these are two separate requirements. No degree of good faith makes it possible for one who fails to meet the notice requirement to qualify as a holder in due course and good faith is not established by showing that a person received an instrument without notice. Of course, these two requirements are closely related so that the conclusion that a holder has failed to satisfy one of these requirements is often bolstered by evidence that he failed to meet the other.

A. NOTICE INSTRUMENT IS OVERDUE

The fact that an instrument is overdue is not, of itself, sufficient to bar a person from being a holder in due course. He must have *notice* of the fact. 3–302(1)(c).

Since a person is charged with notice of what he *has reason to know*, 1–201(25)(c) he is charged with notice of what the instrument itself provides and the date on which he acquires the instrument. If an instrument provides that it is payable on July 1, 1982, the purchaser has reason to know that it is overdue if he takes it on July 2, 1982. This is so even though he can prove that he did not bother to inquire or to examine the instrument to find out when it was due and, in fact, believed that it was not overdue.

Demand instruments present a difficult problem. A person is charged with having notice that a demand instrument is overdue if he has reason to know that he is taking it after demand has been made or more than a reasonable time after its issue. 3–304(3)(c). In the case of *checks* drawn and payable within the states and territories of the United States and the District of Columbia, a reasonable time is presumed to be thirty days. 3–304(3)(c). This presumption is rebuttable and can be overcome by evidence that a longer or shorter period is reasonable.

The Code offers no guidance for determining a reasonable time in the case of demand notes and ordinary demand drafts beyond the general rule that "What is a reasonable time for taking any action depends on the nature, purpose, and circumstances of such action." 1–204(2). The cases are difficult to reconcile if one considers *only* the time periods. Britton, Bills & Notes § 120 (2d ed. 1961). For example, in one case it was held that twenty days was an unreasonable length of time after the issue of a demand note, but in another case it was held that four years was a reasonable time for a similar instrument. In the latter case the court was impressed by the fact that the note was purchased as an investment by one brother from another. Gershman v. Adelman, 103 N.J.Law 284, 135 A. 688

(1927). In an earlier case, it was held that a demand note was not overdue 18 months after its date, primarily because the note was kept alive by continuous payments of monthly interest. McLean v. Bryer, 24 R.I. 599, 54 A. 373 (1903).

The instrument's nature and purpose is significant in determining a "reasonable time after its issue," as shown by the tendency to hold that a reasonable length of time after issue is shorter for a check which bears no interest and does not normally circulate than for a demand draft or note that bears interest and frequently circulates or a certified check that occasionally circulates. Similarly, a reasonable period of time for an interest-bearing certificate of deposit, which commonly is held as an investment, is usually longer than for other kinds of instruments.

When an instrument contains an acceleration clause and the event stipulated occurs, the instrument immediately becomes overdue. This does not, however, prevent the then holder from negotiating the instrument to another. The Code provides that one who purchases such an instrument before the time provided for payment has notice that the instrument is overdue if he has reason to know that the event accelerating the instrument has occurred, but otherwise not. 3–304(3)(b).

Instruments payable in instalments, particularly instalment notes, are quite common. This manner of payment does not interfere with negotiability. When such an instrument contains an acceleration clause, knowledge of a default in the payment of one instalment clearly constitutes notice that the instrument is overdue. Even when such an instrument does not contain an acceleration clause, a person who knows of a default in the payment of any instalment is charged with notice that the instrument is overdue. In fact, he is charged with such notice whenever he has reason to know that any part of the *principal* is overdue. 3–304(3)(a).

Similarly, when instruments issued in a series are payable at different times, a purchaser who has reason to know that one of the instruments is overdue is charged with notice that all are overdue even though they contain no acceleration clause and the instrument he purchases has not yet matured according to its terms. 3–304(3)(a). In contrast, the Code provides that a person can become a holder in due course even though he knows that there has been a default in the payment of *interest* on an instrument. 3–304(4)(f). This difference in treatment is explained by the fact that interest payments frequently are delayed. Comment 6, 3–304.

B. NOTICE INSTRUMENT HAS BEEN DISHONORED

In most cases, a dishonor occurs when a demand for payment or acceptance is properly made upon the party expected to pay, and payment or acceptance is refused or cannot be obtained within the time allowed. 3–507(1)(a). See Chapter 8, Section 2. In the case of bank collections, an instrument is dishonored if it is seasonably returned by the midnight deadline.

3–507(1)(a), 4–301. A dishonor also occurs if presentment is excused and the instrument is not duly accepted or paid. 3–507(1)(b).

Very often, particularly when presented for payment through ordinary banking channels, a dishonored instrument bears a stamp or other notation such as "not sufficient funds," "no account," or "payment stopped." Regardless of actual knowledge, anyone purchasing such an instrument is charged with knowing that the instrument has been dishonored.

Sometimes a dishonored instrument bears no evidence of dishonor, and the person selling such an instrument does not always reveal the dishonor. In these cases, whether or not a person is charged with notice of the dishonor so as to be barred from being a holder in due course depends upon whether or not the trier of fact concludes from other evidence that he had actual knowledge, had received notice, or had reason to know of the dishonor. 1–201(25).

C. NOTICE OF DEFENSE AGAINST INSTRUMENT OR CLAIM TO IT

A purchaser cannot be a holder in due course if he takes an instrument with notice that any person has a defense against it or a claim to it.

In general, a *defense* is asserted negatively as a shield to protect one against being subjected to liability whereas a *claim* is asserted affirmatively as a sword to impose liability on someone else. In the context of the present discussion, a defense might be based on virtually any matter that would furnish a basis for a defense to an action on a simple contract as well as any of the bases that are unique to commercial paper; whereas a claim to the instrument might be asserted by anyone who contends that he has an adverse property interest in it. Thus a claim might be asserted by one who contends that he is the legal or equitable owner or that he has a lien against it. Perhaps the most obvious kind of claim is that which is asserted against a thief or other converter. But the most common kind of claim is that which is asserted by one who is induced to issue or transfer an instrument in a transaction that is voidable in whole or in part because it was induced by fraud, duress, mistake or similar means, or because of incapacity or illegality or for any other reason. Sometimes a claim arises from circumstances that render attempts to transfer void as when they are made by adjudicated incompetents or when they are induced by fraud in the factum that occurs despite the exercise of due care.

Usually, the special facts that give rise to a defense give rise to a corresponding claim and vice versa. Thus, in a common case, a person who is induced to issue or transfer an instrument by fraud might act negatively and assert the defense of fraud as a shield if he is sued on the instrument or he might act affirmatively and assert a claim as a sword to recover the instrument or to impose liability for conversion. In some cases a defense arises without a corresponding claim as where a party issues or transfers an instrument without receiving consideration. Sometimes a claim arises without any corresponding defense; for example, a purchaser of an instrument who pays but is denied delivery has a claim against the seller but he has no need for a defense because he has no liability on the instrument.

In most cases it is obvious whether a party is or is not charged with notice of a claim or defense, but there are a few matters that merit special attention.

1. DISCHARGE

A person has notice of a claim or defense if he has notice that *all* parties have been discharged, but is not barred from being a holder in due course merely because he knows that less than all of the parties have been discharged. 3–304(1)(b), 3–602.

Example 55

> M is induced by fraud to deliver a note to P. P negotiates the note to X. X negotiates the note to Y. Y negotiates the note to H but before doing so strikes out the signatures of X and P thereby discharging them from liability. Here H is charged with notice that P and X are discharged from liability. Since the liability of M and Y has not been discharged, however, H may still be a holder in due course so as to cut off M's defense of fraud.

2. COUNTERCLAIMS AND SET-OFFS

Although a counterclaim or set-off is in one sense defensive, it does not constitute a defense so as to bar one who has notice of it from being a holder in due course.

Example 56

M is induced by fraud to deliver a note for $1000 to P who owes M $400. P negotiates the note to H who knows that P owes M $400 but is unaware of the fraud. H can be a holder in due course and thus free of both the defense of fraud and the counterclaim for $400.

3. Notice Promise of Prior Party Is Unperformed

It is common for a negotiable instrument to be issued or negotiated in return for consideration in the form of a promise. If the promise is broken, this usually gives rise to a claim or defense between the two immediate parties; therefore notice of the breach normally bars a person from being a holder in due course. But a purchaser is not charged with notice of a defense or claim merely because he knows that the instrument was issued or negotiated in return for an executory promise or was accompanied by a separate agreement if he has no notice of a breach. 3–304(4)(b).

4. Instrument as Notice of Defense or Claim

Usually a person will not purchase an instrument until he has examined it. It seems reasonable, therefore, to charge him with having notice of any defenses or claims which are apparent from an examination of the instrument itself. 3–304(1)(a). Some cases are clear. For example, a bungling forgery or an obvious alteration which raises the amount payable, should make a purchaser suspicious and serve as notice. Just how cautious or suspicious a purchaser must be remains a question.

Prior to the Code even though a purchaser received a *complete* instrument he was barred from being a holder in due course if he knew that originally it was blank concerning some material matter. The Code provides that a purchaser is not charged with notice of a defense or claim merely because he knows that an incomplete instrument has been completed unless he also has notice of an improper completion. 3–304(4)(d). Under this provision it is possible for a purchaser to be a holder in due course even though a blank is filled out in his presence if he has no reason to know that the completion is improper. Comment 10, 3–304. For example, in Sun Oil Co. v. Redd Auto Sales, Inc., 339 Mass. 384, 159 N.E.2d 111 (1959), it was held that the plaintiff seller was not prevented from being a holder in due course of a check because its agent received the check from its customer, a gas station operator, knowing that the check had been signed in blank by the defendant, an accommodation party, and that the operator had filled in the plaintiff's name as payee and the amount in the agent's presence according to a common practice among the parties.

The most obvious type of irregularity is an alteration of the amount payable, but an irregularity also might be a change of date, of the payee's name, or of some other material term. The fact that an instrument is antedated or postdated is not, however, such irregularity as to charge a purchaser with notice of a defense or claim, 3–304(4)(a); neither is the fact that early in 1982 an instrument contains an obvious change in the date from January 2, 1981, to January 2, 1982. Comment 2, 3–304. A person who purchases a negotiable instrument is charged with observing what appears on the reverse side of the instrument as well as the face. Consequently, he is expected to heed any irregularities with respect to indorsements. Hence, a person who receives an instrument which has been indorsed on the reverse side, "In payment for poker losses (Signed) J. Smith" would be charged with knowing of a defense in most states even though he did not read the indorsement or know that gambling transactions are illegal.

But the fact that an indorsement is qualified, or for accommodation or restrictive, or contains words of guarantee or assignment, is not sufficient of itself to charge a person with notice of a defense or claim.

5. FIDUCIARIES

Although a person is not charged with notice of a defense or claim merely because he has notice that a person negotiating an instrument is or was a fiduciary, 3–304(4)(e), he is charged with such notice if he knows that a fiduciary had negotiated the instrument in payment of or as security for his own debt or otherwise in breach of duty. 3–304(2). As mentioned earlier, if an instrument is restrictively indorsed for the benefit of the indorser or another, the person who takes the instrument from the restrictive indorsee can be a holder for value, and a holder in due course, only to the extent he applies any value given by him consistently with the restrictive indorsement. But a later holder is not affected by the restrictive indorsement unless, as indicated above, he knows the fiduciary has broken his duty. 3–206(4).

§ 8. Notice of One Defense or Claim Subjects Holder to All

The principle that a purchaser cannot be a holder in due course if he takes an instrument with notice of any defense against it or claim to it by any person, 3–202(1)(c), applies even though the defense or claim of which he has notice does not affect the defense or claim ultimately asserted against him.

Example 57

M is induced by *fraud* to deliver a note to P. P negotiates the note to X. X is induced by *duress* to negotiate the note to Y. Before maturity Y negotiates the note to H for value. H is aware of the duress but unaware of the fraud. Since duress is a defense, H is not a holder in due course. If H sues M, P's fraud is a good defense even though H was unaware of the fraud when he purchased the note. Similarly, if H had been aware of the fraud but was not aware of the duress, he would not have been a holder in due course and consequently would have been required to return the note to X if X had asserted a claim to it on the basis of the duress.

§ 9. Payee as Holder in Due Course

Under the Code, a payee, like any other holder, can be a holder in due course so long as he meets the usual requirements. This is so regardless of whether he takes the instrument by issue (first delivery) from the maker or drawer or from a third party (remitter).

Comment 2, 3–302. In fact, in most cases where he acts in good faith and gives *executed* consideration (value) the payee is a holder in due course because normally he cannot be charged with notice that the instrument is overdue or has been dishonored or is subject to any claim or defense. If the issuing party happens to have a defense of which the payee is not aware—for example, material mutual mistake—the existence of the defense does not prevent the payee from being a holder in due course. But recognizing the payee as a holder in due course would do him no good because even a holder in due course is subject to any defense of the party with whom he deals. 3–305(2). About the only case in which it makes any difference whether a payee is a holder in due course is that in which he does not deal directly with the maker or drawer but instead obtains the instrument from a remitter who obtained it from the maker or drawer. In this case, if the payee qualifies as a holder in due course, he takes the instrument free of any defense based on the remitter's wrongdoing; otherwise not.

Example 58

X orders goods from P on credit, but P refuses to sell to X on this basis. Therefore, X induces his bank to sell to him, on credit, a draft drawn by the bank on itself naming P as payee. X then forwards the draft to P. If P ships the goods, acts in good faith, and meets the requirement with respect to notice, he is a holder in due course. Consequently, if P demands payment from the bank and the latter refuses payment on the ground that X induced it to issue the draft by fraud, this defense will not be effective against P. The result and analysis would be the same if it is as-

§ 10. Taking Through Holder in Due Course—The Shelter Provision

Although it is important to keep in mind the requirements for being a holder in due course it is almost as important to remember that under the *Shelter Provision* of Section 3–201(1), the "transfer of an instrument vests in the transferee such rights as the transferor had therein." By virtue of this broad principle (which applies to most other forms of property as well), even though a person does not satisfy the requirements for being a holder in due course, he is entitled to enjoy all of the benefits of that status if he can prove that someone prior to himself in the chain of ownership was a holder in due course. The primary significance of the Shelter Provision and the basis for its name is the fact that it enables one who is not a holder in due course to share the shelter from claims and defenses enjoyed by a holder in due course through or from whom he acquired the instrument.

Example 59

M is induced by P's fraud to execute and deliver a note. P negotiates the note to H under circumstances making H a holder in due course. After maturity, H negotiates the note to T who knows it is overdue. T is entitled to recover the amount of the note from M. Even though T is not a holder in due course because he knew the note was overdue when he received it, he is free of M's defense because he has all of the rights of H who was a holder in due course.

Giving the transferee who is not a holder in due course the rights of his transferor who is, has the effect of increasing the market for instruments held by holders in due course. But this rule does not apply to improve the position of a transferee who has been a party to any fraud or illegality affecting the instrument or who as a prior holder had notice of a claim or defense. For example, a payee who has induced the maker to issue a note by fraud cannot wash the paper clean by running it through a holder in due course and then purchasing it. And, of course, one to whom a holder in due course transfers an instrument as security, acquires no greater interest in the instrument than the amount due on the obligation secured.

Example 60
HDC holds a note made by M for $15,000. HDC borrows $10,000 from L and negotiates the note for $15,000 to L as security for the loan. The interest of L in the note made by M never exceeds the amount due on the $10,000 loan.

§ 11. Transactions that Cannot Give Rise to Holder in Due Course

Even though he satisfies the usual requirements previously described, a person cannot become a holder in due course of an instrument: "(a) by purchase of it at judicial sale or by taking it under legal process; or (b) by acquiring it in taking over an estate; or (c) by purchasing it as part of a bulk transaction not in the regular course of business of the transferor." 3–302(3). In these situations it is normally understood that the transferee merely is acquiring the rights of the prior holder and there is no substantial interest in facilitating commercial transactions, which is the underlying reason for giving holders in due course their special advantages. Of course, in accordance with the Shelter Provision discussed in the preceding paragraph, if the prior holder has the rights of a holder in due course, the transferee under one of the above transactions acquires the same rights. 3-201(1).

§ 12. Assignee with Principal Advantage of Holder in Due Course

In order to make it possible for the assignee of rights that arise from a contract that is not negotiable to acquire the principal advantage that normally is reserved for a holder in due course, Section 9-206(1) provides:

Subject to any statute or decision which establishes a different rule for buyers or lessees of consumer goods, an agreement by a buyer or lessee that he will not assert against an assignee any claim or defense which he may have against the seller or lessor is enforceable by an assignee who takes his assignment for value, in good faith and without notice of a claim or defense, except as to defenses of a type which may be asserted against a holder in due course of a negotiable instrument. (Emphasis added.)

The italicized portion of Section 9-206(1) above preserves for buyers and lessees of consumer goods the benefits of decisions rendered and statutes enacted in many states to protect such persons against the risk of losing defenses they had against persons with whom they dealt in the underlying business transaction, either by expressly waiving in the original contract the right to assert such defenses against assignees, as provided by Section 9–206(1), or by executing negotiable instruments which might later be acquired by holders in due course. (See Chapter 10, Section 14.)

§ 13. Collecting Bank as Holder in Due Course

Suppose that P receives a check for $5,000 in the mail as a gift from D. The check is drawn on D's bank, called the Payor Bank in 4–105(b). P, under the right circumstances, could personally present the check to Payor Bank and demand payment. But whether the item is paid or dishonored, Payor Bank does not become a holder. The item is either paid and returned to D or dishonored and returned to P. If Payor Bank's actions are allegedly improper, the legal issues are resolved without reference to the holder in due course doctrine.

Suppose, however, that P deposits the check for collection in his checking account. The check is delivered to his bank, called the Depositary Bank in 4–105(a), with a restrictive indorsement, "for deposit only." 3–205(c). The Depositary Bank will then attempt to collect the item, i. e., will act as P's agent in presenting the item for payment to Payor Bank. This might be done through a local or regional clearing house or with the help of intermediary banks and the federal reserve system. In either case, Depositary Bank is also a Collecting Bank under 4–105(d). As such, Collecting Bank is a holder but must "pay or apply any value given * * * for or on the security of the instrument consistently with the indorsement * * *." 3–206(3). To the extent that Collecting Bank does so, it becomes

Ch. 10 *HOLDERS IN DUE COURSE*

a holder for value. To the extent that Collecting Bank otherwise complies with Section 3–302 it becomes a holder in due course. 3–206(3).

In the normal course of events, Collecting Bank will not become a holder in due course. If the item is paid by Payor Bank, Collecting Bank will make a final settlement for the amount to the account of P, its Customer. If the item is dishonored, it will be returned by Collecting Bank to P. In either case, Collecting Bank has not given value. But suppose P is a valued customer and Collecting Bank gives him a "provisional" settlement. P is permitted to draw against the check before it is paid by Payor Bank with the understanding that if it is dishonored Collecting Bank has a right of "charge-back." See 4–201(1). Suppose, further, that P immediately withdrew $5,000 against the provisional credit. At this point, Collecting Bank has a security interest in the item to the extent of $5,000, 4–208(1)(a), and this constitutes value for determining whether Collecting Bank is a holder in due course. Thus, *if* the check were dishonored by Payor Bank and *if* P became insolvent, Collecting Bank would be a holder in due course of the check against D, the drawer. In this situation, the defense of "no consideration" would not be available to D. See 3–408, 3–305(2).

§ 14. Special Treatment of Consumers

As explained above, there are two basic methods by which a transferee can acquire contract rights free from defenses that were good against his transferor or some prior party. First, he might acquire a negotiable instrument as a holder in due course. Second, he might take an assignment of a non-negotiable contract in which the obligor has agreed not to assert defenses or claims against an assignee. The existence of these means has greatly facilitated a vast amount of credit which in turn has contributed greatly to the economy. Businessmen, lured by freedom from possible defenses, have received such paper in payment for goods or services, by purchasing it, and by taking it as security for loans.

In normal commercial settings, these devices for cutting off defenses have been at least benign and at best beneficial. When applied, however, by two professionals with a non-professional in the middle, these devices often result in grave injustice, especially when the non-professional is uninformed, unsophisticated, and economically disadvantaged, as he too frequently is.

The unfairness has been most striking in cases in which a consumer—who may be defined generally as one who purchases or leases goods or services primarily for personal, family or home use—executes a negotiable instrument, most frequently an instalment note, or a contract waiving defenses against an assignee, but does not receive what he bargained for. When he refuses to make further payments, he is sued by the person to whom the seller's rights have been transferred, usually a financial institution, and then learns that the non-performance of the seller with whom he contracted is not a good defense against its transferee. After paying the financial institution in full, the consumer must then pursue expensive remedies against a defaulting contractor who in many cases, is insolvent.

As applied in these and similar cases, the devices for cutting off defenses have been among the natural targets of the consumer movement. In recent years they have been sharply limited or prohibited by some state courts and legislatures and have been declared to be an unfair trade practice by the Federal Trade Commission. Some examples of this pattern of regulation will be mentioned briefly.

A. THE COURTS

In most cases in which courts have refused to give effect to the devices intended to limit the defenses of a consumer sued by financial institutions, the decisions have been based on the terms of Section 3-302, which defines a holder in due course, or the terms of Section 9-206 which declares the legal boundaries of consumer agreements to refrain from asserting defenses against assignees.

One might reasonably suppose that since the financial institution is in a close relationship with the dealer, it is most likely to be denied protection from defenses on the ground that it had notice of a defense, which constitutes a bar to such protection under both Section 3-302 and Section 9-206. This is especially so since Section 1-201(25) charges a person with notice of a defense if, "from all of the facts and circumstances known to him * * * he has reason to know that it exists." But very few cases that have ruled against financial institutions have done so on this basis, presumably because the defense must be known at the time that the instrument is taken. Instead, in the bulk of such decisions, the courts have expressly or impliedly relied on one or both of two related bases: (1) lack of good faith, and (2) the unity of the dealer and the financial institution—and they usually have supported these bases by one or more policy arguments.

1. LACK OF GOOD FAITH

In reviving seemingly lost defenses that consumers have asserted against financial institutions to whom consumer notes and contracts have been transferred, many courts have relied on the fact that neither Section 3–302 nor Section 9–206 offers the advantage of cutting off defenses to one who does not take the instrument in good faith in the sense of honesty in fact. 1–201(19). Among the reasons that courts have expressly or impliedly given for holding that a financial institution has not met the good faith requirement are the following: it acted in collusion with the dealer to defraud the consumer; it had reason to be suspicious because it was familiar with the dealer's practices or the practices in the dealer's industry, or because the price at which the dealer was willing to sell was extremely low; its failure to inquire seemed to be motivated by a fear of learning the truth; its speed in purchasing the paper suggested a primary purpose to cut off the consumer's defenses, rather than a normal business purpose; its legal separation from the dealer was only a guise intended to cut off the consumer's defenses; or it failed to observe the reasonable commercial standards of its business—a test expressly included in the Section 3–302(1)(a) requirement of good faith in the 1952 Official Text of the Code but omitted from all later Official Texts. For an example, see General Inv. Corp. v. Angelini, 58 N.J. 396, 278 A.2d 193 (1971).

2. UNITY OF DEALER AND FINANCIAL INSTITUTION

Many courts that have refused to enforce devices intended to cut off consumer defenses have relied on the fact that Section 3–305(2), which declares what defenses are good against a holder in due course, makes it clear that even a holder in due course does not take an instrument free from defenses of the person with whom he deals; and they also have relied on the implication of Section 9–206 that the assignor and the assignee must be acting separately. Freedom from defenses does not extend under either of these sections to cases in which the court finds that the dealer and the financial institution were acting together as a single entity. Among the reasons expressly or impliedly given by courts for finding a financial institution to be in unity with a dealer so as to subject it to consumer defenses that otherwise would be barred are the following: it participated in the underlying transaction by furnishing forms, investigating credit, approving the consumer, counseling the dealer and in similar ways; it dealt directly with the consumer; it was only an agent for the dealer; it had a substantial voice in controlling the underlying transaction; it was but a part of the same corporate family, holding a controlling interest in the dealer's stock or vice versa; it was named in the original printed forms; it, along with the dealer, was named as a "seller" in the contract signed by the consumer; and its speed in acquiring the consumer paper indicated that there was in fact but one transaction, with the consumer on one side and the dealer and financial institution on the other. For useful discussion, see Note, *The Close Connectedness Doctrine: Preserving Consumer Rights in Credit Transactions*, 33 Ark.L.Rev. 490 (1980).

3. SUPPORTING ARGUMENTS

In most cases in which a court has denied a financial institution freedom from defenses on the ground of lack of good faith or unity of action with the dealer, despite the presence of a defense-defeating device, the court has bolstered its decision with one or more of the following arguments: a financial institution that participates closely in the original transaction is not within the class of persons intended to benefit from the holder in due course concept; a financial institution that furnishes forms and procedures designed to put the consumer in a straitjacket should not be permitted to benefit from such conduct; the free flow of credit basis for the holder in due course concept has no application to consumer financing which normally involves only the consumer, a dealer and a finance company; the holder in due course concept was designed to facilitate the free flow of commercial paper among legitimate businesses and was not intended to permit a transferee to insulate himself from fraudulent practices by feigning ignorance; stricter standards for those who wish to take commercial paper free from defenses encourages them to police dealers more carefully and to ferret out the dealers who are dishonest or financially unstable; in the light of the totality of the transaction, it would be unconscionable to deny the consumer his defenses; the Code itself by Section 2–302 encourages the courts to refuse to carry out unconscionable provisions in sales transactions; and the Code encourages a general policy of protecting consumers by providing in Section 9–206(1) that waiver of defense clauses are "subject to any statute or decision which establishes a different rule for buyers or lessees of consumer goods." See, e. g., Unico v. Owen, 50 N.J. 101, 232 A.2d 405 (1967); Rosenthal, *Negotiability—Who Needs It*, 71 Colum.L.Rev. 375 (1971).

B. THE LEGISLATURES: UNIFORM CONSUMER CREDIT CODE

The difficulty courts have had in trying to strike a proper balance between the policy of promoting the free transfer of commercial paper and the policy of protecting the consumer who buys on credit has led legislatures in many states to deal with the matter by statutes which limit the operation of devices intended to bar consumers from asserting defenses. Some statutes have focused their attacks almost entirely on devices intended to cut off defenses of consumers while others, such as the Uniform Consumer Credit Code, have dealt with these and numerous other aspects of consumer credit as well.

Most such statutes that have been adopted or proposed have defined a consumer transaction as one wherein goods or services intended primarily for personal, family, or household use, are sold or leased; or in some similar way. They have then limited defense-defeating devices in one or more of the following ways: by forbidding the use of negotiable notes in consumer credit transactions; by flatly providing that notes given in consumer credit transactions are not negotiable; by denying holder in due course status to one who knowingly purchases consumer paper; by requiring that consumer notes be so designated and then providing that that such a note cannot be negotiable; by voiding agreements by consumers to waive defenses against transferees; and by providing that the transferee of a consumer note or contract is subject to any defense of which he receives notice within a stated period (typically one month) after the consumer receives notice of the transfer, notwithstanding any agreement by the consumer to the contrary. In addition, some statutes contain special provisions re-affirming the courts' power to refuse to enforce unconscionable provisions against consumers.

The 1974 version of the Uniform Consumer Credit Code takes a direct approach to this problem. Section 3.307 provides that in a consumer credit sale or lease, the creditor "may not take a negotiable instrument other than a check dated not later than ten days after its issuance as evidence of the obligation of the consumer." Section 3.404(1) provides that an assignee of the rights of a seller or lessor in the consumer transaction is subject to "all claims and defenses" arising from the transaction "notwithstanding that the assignee is a holder in due course of a negotiable instrument issued in violation of" Section 3.307. Section 3.404(4) then provides that an "agreement may not limit or waive the claims or defenses of a consumer under this section." Provided that the consumer makes a good faith effort first to obtain satisfaction from the seller or lessor and gives proper notice, UCCC 3.404(2), these sections eliminate the shield previously provided by the holder in due course doctrine and "waiver of defense" clauses in any case where the financial institution purchases the instrument or agreement from the seller or lessor. In addition, the aggrieved consumer may recover a penalty of "not less than $100 nor more than $1,000" from a creditor who has violated the Code. But the scope of the claim or defense may not exceed the "amount owing to the assignee with respect to the sale or lease of the property or services as to which the claim or defense arose at the time the assignee has notice of the claim or defense." UCCC 3.404(2).

Suppose, however, that L, a lender, lends M $1,000 and takes a negotiable promissory note for that amount. M, with L's blessing, uses the $1,000 to purchase goods from S which later turn out to be defective. M has paid S cash and S is now insolvent. May M assert his defense and claim against S against L or his transferee on the instrument? Under the UCCC, the answer is no unless the loan was "conditioned" upon M's purchase of the goods from "the particular seller." UCCC 3.405(1)(e). It is not enough that L knowingly made a purchase money loan. Other knowledge by L or connections with S may, however, subject L to M's claims and defenses. For example, L assumes the risk if he knows that S "arranged for the extension of credit by the lender for a commission" or "directly supplies the seller * * * with the contract document used by the consumer to evidence the loan, and the seller * * * has knowledge of the credit terms and participates in preparation of the document." UCCC 3.405(1). Other factors are relevant in this complex analysis, and the statutes and case law in your state must be consulted. (At this writing, some version of the UCCC has been adopted in ten states.)

C. FEDERAL TRADE COMMISSION

In 1975, the Federal Trade Commission promulgated regulations designed to deal with the holder in due course problem in consumer sales or leases in interstate commerce. See 16 C.F.R., Part 433. The regulations cover two basic transactions: 1) the seller or lessor of goods or services takes or receives a consumer credit contract from the consumer; and 2) the seller or lessor of goods or services accepts, as full or partial payment, the proceeds of any purchase money loan, which is defined as a cash advance for a finance charge which is "applied, in whole or in substantial part, to a purchase of goods or services from a seller who (1) refers consumers to the creditor or (2) is affiliated with the creditor by common control, contract, or business arrangement." These transactions are identical to the direct and indirect extensions of credit regulated by the Uniform Consumer Credit Code and other state legislation. Within the scope of these transactions, the Regulations provide that it is "an unfair or deceptive act or practice" within the meaning of Section 5 of the Federal Trade Commission Act for the seller or lessor to fail to have included in either the consumer credit contract taken from the consumer or the consumer credit contract made in connection with the purchase money loan the following provision in at least ten point, bold face type:

NOTICE

ANY HOLDER OF THIS CONSUMER CREDIT CONTRACT IS SUBJECT TO ALL CLAIMS AND DEFENSES WHICH THE DEBTOR COULD ASSERT AGAINST THE SELLER OF GOODS OR SERVICES OBTAINED PURSUANT HERETO OR WITH THE PROCEEDS HEREOF. RECOVERY HEREUNDER BY THE DEBTOR SHALL NOT EXCEED AMOUNTS PAID BY THE DEBTOR HEREUNDER

This notice on the face of the note would clearly impair its negotiability under state law.

As federal law, these regulations arguably preempt state law to the extent that lesser protection is provided. Similarly, if state law provides greater protection, then it should apply to that extent. This mesh of state and federal law would be critical in the following case:

Example 61

S, in interstate commerce, sold B, a consumer, goods on credit and took a consumer credit contract *without* the NOTICE required by the FTC regulations. The contract contained a clause waiving all claims and defenses and was accompanied by a negotiable promissory note. S transferred the contract and negotiated the note to H, a holder in due course under the Uniform Commercial Code. Thereafter, S failed to deliver the goods and became insolvent. H sued B on the contract and note and B raised the defense of failure of consideration. (1) The defense is not available under Section 3–305 of the UCC. (2) The seller's failure to include the NOTICE is an unfair trade practice under the FTC Act, but enforcement of those regulations is by public officials rather than private parties and the regulations do not specify the legal consequences if the NOTICE is omitted and the contract transferred to a holder in due course. (3) The defense is available under the UCCC to the full amount of the unpaid contract price. Presumably, state law (the UCCC) with its broader consumer protection would be available to supplement federal law. But if the UCCC were not enacted and H was a HDC under the UCC, B, presumably, would have to pay.

In sum, the question "who needs negotiability in consumer credit transaction" has been answered by the FTC, the UCCC and many state legislatures: "no one." The doctrine should be abolished. What justifies this answer? The justification comes from a combination of fairness and efficiency. It is unfair to place upon the individual consumer full liability to the holder for defective products or services and the responsibility for trying to recover over against the seller or lessor. More importantly to some, it may be inefficient to do so. Arguably, it is less costly for the financial institution to assess and manage the risk of doing business with the seller or lessor than the consumer. The professional can assess the risk and discount the transfer of paper with greater effectiveness than if the various cut-off devices were preserved. Thus, the risk of the seller's defective performance is managed in the transaction of transfer and the consumer, by being allowed to assert defenses, is spared the anguish and frustration of fighting both the fiancial institution and the seller.

CHAPTER 11
DEFENSES AND CLAIMS

Although the principal advantage of being a holder in due course is freedom from defenses and claims, it is important to recognize that this immunity has its limits. The present chapter considers the various defenses and claims and the extent to which they may be asserted effectively against those who have the rights of a holder in due course and those who do not. (For the distinction between a defense and a claim see pages 258–259)

§ 1. Defenses Effective Against Holder in Due Course

For purposes of analysis, it is convenient to group defenses which can be asserted against a holder in due course into two general categories. First, there are defenses which arise from the transaction by which he acquires the instrument or at some later time. Second, there are defenses which arise before he acquires the instrument.

A. DEFENSES ARISING AT TIME OF TRANSFER OR LATER

The need for good faith and other requirements for becoming a holder in due course minimize the likelihood of a defense arising from the transaction of negotiation to a holder in due course. Occasionally, however, a defense arises at that time. For example, as

Ch. 11 *DEFENSES AND CLAIMS*

part of the transaction of transfer, the holder in due course may agree to cancel his transferor's obligation by striking out his signature; or unknown to the holder in due course, a transaction may be voidable on the ground of illegality; or the parties may be acting on the basis of a mutual material mistake. Also, defenses may arise after he acquires an instrument. For example, an obligor might pay the holder in due course; the latter might renounce his rights in a separate writing without giving up the instrument; the transferor or a prior party might obtain a discharge in bankruptcy; or a secondary party might be discharged by the holder's failure to make a proper presentment or to give due notice of dishonor. *Regardless of the nature of the defense arising when he receives the instrument or later, a holder in due course is subject to the defense just as if he were not a holder in due course.*

B. DEFENSES ARISING BEFORE NEGOTIATION TO HOLDER IN DUE COURSE

Many defenses, including those already mentioned, may arise *before* the time the instrument is negotiated to a holder in due course. The principal advantage of being a holder in due course is that most of these defenses are cut off when he acquires it. The defenses which arise prior to the time the instrument is negotiated to a holder in due course sometimes are classified as *real* and *personal*. Real defenses are good against holders in due course, as well as against persons who are not, and personal defenses are good only against those who do not have the rights of a holder in due course. Although the Code does not use the terms real and personal as applied to defenses, many courts and lawyers continue to do so.

The defenses available to a prior party to the instrument and not cut off when the instrument is negotiated to a holder in due course are set forth in Section 3–305(2) of the Code which provides:

> To the extent that a holder is a holder in due course he takes the instrument free from * * * all defenses of any party to the instrument with whom the holder has not dealt except
>
> (a) infancy, to the extent that it is a defense to a simple contract; and
>
> (b) such other incapacity, or duress, or illegality of the transaction, as renders the obligation of the party a nullity; and
>
> (c) such misrepresentation as has induced the party to sign the instrument with neither knowledge nor reasonable opportunity to obtain knowledge of its character or its essential terms; and
>
> (d) discharge in insolvency proceedings; and
>
> (e) any other discharge of which the holder has notice when he takes the instrument.

1. INFANCY

Infancy is a defense against a holder in due course to the same extent that it is a defense to an action on a simple contract under the state law governing the transaction. With very few exceptions—for example, contracts of enlistment, marriage, and those specially provided by statute—the contracts of an infant are voidable entitling the infant to disaffirm or rescind his contract even though he has received and enjoyed the benefits. In most states the same applies to contracts for necessaries; but if an infant rescinds a contract for necessaries received, he is nonetheless bound to pay their fair value on the theory of quasi contract. Consequently, in the great majority of transactions an infant who incurs liability on a negotiable instrument may avoid this liability in whole or in part even against a holder in due course. See Trenton Trust Co. v. Western Sur. Co., 599 S.W.2d 481 (Mo.1980).

2. OTHER INCAPACITY

Incapacity also may be recognized if at the time the transaction occurred the defendant was insane, intoxicated or a corporation acting outside the powers given by its charter, and in a few other special circumstances. Unlike infancy, which enables a defendant to avoid liability against a holder in due course even though his liability is merely *voidable*, parties limited by other types of incapacity are shielded from such liability only if the governing state law declares that such incapacity renders transactions *void*. Since incapacity other than adjudicated incompetency normally renders transactions *voidable* and not void, incapacity other than infancy and adjudicated incompetency usually does not constitute a valid defense against holders in due course.

3. DURESS

Duress occurs when one person, by the exercise of wrongful pressure, induces another to become fearful and to do something he otherwise would not do. The wrongful pressure may consist of physical violence, but usually it consists of threats. A threat to start a civil action usually is not considered to be wrongful. The threat to have someone prosecuted for a crime, though not wrongful in itself, is considered to be wrongful when made to induce someone to enter a contract. Normally duress merely renders a contract voidable and so is not a good defense against one having the rights of a holder in due course. In its very extreme form, however, as where one party points a loaded gun at another or physically compels him to manifest his assent to contract, duress renders a transaction void. Such duress would be a good defense even against one having the rights of a holder in due course. Actually, duress of this kind is rare.

4. ILLEGALITY

In general, a transaction is illegal if it is contrary to public policy as provided by statute or as declared by the courts. A negotiable instrument may be issued or negotiated in a transaction which is illegal because it involves gambling, usury, bribery, concealing a crime, promoting immorality, restraining trade; because it occurs on Sunday; or for many other reasons. Generally, the defense of illegality is not good against a holder in due course even though the instrument in question was issued or negotiated in direct violation of a statute or was given for a consideration held to be against public policy. However, if the courts of a state hold, either on the basis of statutory construction or otherwise, that a particular transaction is *void*, a party who issues, accepts, or negotiates an instrument as part of the transaction can effectively defend on the ground of illegality even against a holder in due course. See Sandler v. Eighth Judicial District Court, 96 Nev. 164, 614 P.2d 10 (1980) (check issued for purpose of gambling void, illegality a defense against HDC).

5. Fraud

The Code recognizes that a person might be induced to sign a negotiable instrument by either of two quite different kinds of fraud. The first kind, sometimes called *fraud in the inducement* or *fraud in the procurement*, occurs when the defrauded person is fully aware of the character and essential terms of the instrument he signs but is induced to want to sign by deception with respect to some other matter. For example, a person might be induced to sign and deliver his check for $1,000 in payment for a stone which the seller payee has fraudulently represented to be a diamond. The second kind of fraud, sometimes called *fraud in the factum* or *fraud in the essence*, occurs when the defrauded person is induced to sign an instrument by deception with respect to its *character or terms*. For example, he might be induced to sign a note for $10,000 relying on a false representation that it is only a receipt or that the amount is only $1,000. Fraud of the *first* kind is *not* a defense against a holder in due course. The *second* kind of fraud *might or might not* be a good defense, depending on whether the defrauded party can prove that *he had no reasonable opportunity to learn the true character or terms of the instrument.* If he can prove this, fraud regarding the character or terms of the instrument is a good defense even against a holder in due course; otherwise not. To determine whether the deceived party had a reasonable opportunity to avoid the deception, a court may consider any relevant factors, including the age, mental and physical condition, education and experience of the defrauded party and, if he was victimized because he trusted the other party, his reasons for the trust. For example, a person deceived by clever sleight of hand by an honest-appearing defrauder or who was deceived because he trusted a theretofore trustworthy son, would have a better chance of asserting the fraud successfully against a holder in due course than one who was deceived because he trusted a stranger who acted suspiciously.

Example 62

After negotiations, B, the purchasing agent of Corporation, agreed to purchase 10 water coolers from S for installation in the plant. At the closing of the deal in B's office, S, the sales manager of Seller, produced two documents for B's signature. The first was a four page "contract for sale." B read the page containing the negotiated terms and signed without reading three pages of "fine print." The second was a one page writing with some blanks that had been filled in and three paragraphs of fine print. S stated to B: "This is the standard bond securing the water coolers against damage before they are paid for." B signed the writing without comment and without reading it. In fact, B had signed a negotiable note and an agreement creating an Article 9 security interest in the water cooler. Later, Seller negotiated the note and security interest to H, a holder in due course. Shortly thereafter, defects were discovered in the water coolers and Seller was unable to repair or replace them. In a suit by HDC against Corporation, Corporation argued that a misrepresentation as to the character of the writing had been made and that HDC was subject to the defense under Section 3–305(2)(c).

Corporation will probably lose. Even though B was induced by the misrepresentation to sign the writing without "knowledge" of its character or essential terms, the question is whether he was deprived of a "reasonable opportunity" to obtain that knowledge. He was not prevented from reading it—he chose not to. Given B's experience, intelligence and the business setting, he was capable of reading and understanding the document and was not reasonable in relying exclusively upon S' characterization. See Exchange International Leasing Corp. v. Consolidated Business Forms Co., Inc., 462 F.Supp. 626 (W.D.Pa.1978).

6. DISCHARGE IN INSOLVENCY PROCEEDINGS

The main purposes of bankruptcy and other insolvency proceedings—to restore the debtor's motivation and to give society the benefit of his services—would be frustrated if debtors could not be completely discharged from their liability on negotiable instruments. Consequently, if a debt that is evidenced by a negotiable instrument is discharged in an insolvency proceeding, the discharge is a good defense even against a holder in due course, regardless of notice. 3–305(2)(d).

7. DISCHARGE OF WHICH HOLDER IN DUE COURSE HAS NOTICE

As stated earlier (page 260,) a holder is not barred from being a holder in due course by having notice when he takes an instrument that one or more parties to it have been discharged, unless he has notice that *all* of the parties to it have been discharged. If the holder in due course takes an instrument with notice that any party to it has been discharged, however, that discharge is a good defense against him. If he does not have such notice, the discharge usually is not a good defense. There are a few exceptions based on statutes that supersede Article 3. For example, as mentioned above, a discharge in an insolvency proceeding usually is a good defense against a holder in due course regardless of whether he has notice of it when he takes the instrument.

All of the classes of defenses that have been described above as not being cut off when the instrument is acquired by a holder in due course are mentioned in Section 3–305(2). In addition to the defenses set forth in Section 3–305(2) there are several other defenses that may not be cut off when the instrument is acquired by a holder in due course.

8. MATERIAL ALTERATION

The first of these defenses is material alteration which was discussed earlier. (See Chapter 7, Section 2(G)). A material alteration in the nature of unauthorized completion of an incomplete instrument is not, as seen earlier, a defense against a holder in due course.

But a material alteration in the nature of a deletion, change, or addition to a completed instrument normally can be asserted against a holder in due course who is entitled to enforce the instrument only according to its original tenor. If the negligence of the person who is asserting the alteration as a defense substantially contributed to the alteration, however, he is precluded from asserting the alteration against a holder in due course. 3–406.

9. UNAUTHORIZED SIGNATURE INCLUDING FORGERY

Thus far, the discussion of defenses that might be effective against a holder in due course has been limited to defenses that might be asserted by a party to the instrument; that is, someone who is liable on the theory that he signed the instrument. But when a holder in due course sues a person on the theory that he signed the instrument, that person can defend by proving even against the holder in due course that his signature was unauthorized unless, of course, he is precluded from proving it by the fact that his negligence substantially contributed to the unauthorized signature. 3–406. Observe that this problem will rarely arise if the unauthorized signature is that of an indorser in the chain of title. The reason is that if there is an unauthorized signature in the chain of title there normally can be no later holder in due course, because there can be no later *holder* unless the instrument is later acquired by the person whose signature was unauthorized.

§ 2. Defenses Against One who Lacks the Rights of a Holder in Due Course

One who lacks the rights of a holder in due course is subject to any defense of the party with whom he dealt or of any prior party. Thus, he is subject to all of the defenses that are good against a holder in due course and all of the other defenses that might be asserted in an action on a simple contract, 3–306(b), including lack or failure of consideration, non-delivery (where parties agreed that delivery would be required), incapacity, illegality, fraud, duress, undue influence, material mistake, and many others. In general, these defenses are established by the same evidence when asserted by a person who is sued on a negotiable instrument as when asserted by one who is sued on a simple contract.

Even though consideration rarely presents a serious problem, it merits brief discussion. In general, "want or failure of consideration is a defense as against any person not having the rights of a holder in due course." 3–408. Although consideration is used in the same sense as in the law of simple contracts, in the law of commercial paper there are several special situations wherein consideration is not required to render a promise enforceable.

Thus, Section 3–408 provides that "no consideration is necessary for an instrument or obligation thereon given in payment of or as security for an antecedent obligation of any kind." Under this provision, if a person signs an instrument intending that it be used as security for an antecedent obligation, whether his own or a third party's, he is liable in the capacity in which he signs even though the creditor does not furnish any consideration by extending the time of payment or by incurring any other detriment in return. Comment 2, 3–408. According to Comment 2 of Section 3–408, this provision, "is intended also to mean that an instrument given for more or less than the amount of a liquidated obligation does not fail by reason of the common law rule that an obligation for a lesser liquidated amount cannot be consideration for the surrender of a greater." This reading of Section 3–408 has not been accepted by everyone. It is clear, however, that consideration is not required to render a cancellation or renunciation of liability on an instrument binding. (See pages 325–327). Also, if an accommodation party signs an instrument after value has been given for it, he becomes liable in the capacity in which he signs, to the holder at the time or to any later holder, even though the holder incurs no legal detriment in return for the signing. Comment 3, 3–415.

In addition to these special cases wherein the Code renders promises enforceable even though not supported by consideration, the Code follows the law of simple contracts by recognizing that no consideration is necessary if the doctrine of promissory estoppel or some other equivalent of consideration obviates the need for it. 1–103. Also, a seal on a negotiable instrument has its normal effects so long as they are consistent with the provisions of Article 3. Comment, 3–113. Since there is nothing in Article 3 to the contrary, instruments under seal are governed by any state law by which a seal either creates a presumption of consideration or gives rise to a longer statute of limitations.

The nondelivery of an instrument—even one that is incomplete—is not a defense against a holder in due course, nor is the fact that an instrument was delivered for a special purpose or is subject to some condition which has not been satisfied. Any of these defenses, however, is good against one who lacks the rights of a holder in due course. 3–306(c). This is so regardless of whether the defense is asserted by a maker, drawer, acceptor, or indorser.

A. OUTSIDE AGREEMENT

A person sued on a negotiable instrument might assert a defense based on an outside writing. A negotiable instrument might be but one of several writings that are executed as part of the same transaction. An outside writing might contain terms intended to control the rights and duties that arise from the instrument. Often these terms are not consistent with the normal effect of the instrument and limit the duties of the party signing the instrument. If the negotiable instrument is later acquired by a holder in due course, he and his transferees will not be adversely affected by any term of the outside writing unless he had notice of the term when he took the instrument. But if the instrument is never acquired by a holder in due course, any transferee holds the instrument subject to the terms of the outside writing executed as part of the same transaction that gave rise to the instrument whether he knew of the outside writing or not. For example, if a contemporaneous outside agreement provides that a negotiable instrument is to be discharged without need for payment on the occurrence of some specified event, one lacking the rights of a holder in due course would be bound by this.

It should be noted that the parol evidence rule does not bar a defendant from proving a contemporaneous outside agreement even though it is inconsistent with the terms of the negotiable instrument, because the rule does not apply at all unless the instrument in question is intended to be the complete and final embodiment of the agreement of the parties and it is assumed in this case that the parties intended otherwise. Furthermore, it is a basic rule that writings executed as part of the same agreement must be read together in determining the rights and duties of the parties. Perhaps it is also noteworthy that the outside writing does not interfere with the negotiability of the instrument because negotiability is determined by the terms of the instrument itself and not by anything said or written outside the instrument. See Chapter 5, Section 4.

In addition to being subject to many other possible defenses, one who lacks the rights of a holder in due course is subject to the defense of *discharge* regardless of the basis for the discharge or his lack of knowledge when he acquires the instrument. This applies to all forms of discharge, including those which might operate as a discharge on a simple contract. For example, if a negotiable instrument is issued in return for the payee's promise to render personal services, the payee's death prior to performance would discharge the maker's or drawer's obligation just as it would discharge an obligation on a simple contract. Because of its importance and pervasive nature, the subject of discharge will be discussed separately in Chapter 12 where a number of bases for discharge peculiar to the law of negotiable instruments will be considered.

B. DEFENSE THAT THIRD PARTY HAS CLAIM—JUS TERTII

Sometimes a party sued on a negotiable instrument has no defense of his own but knows that some party prior to the plaintiff has a claim that would entitle him to recover the instrument from the plaintiff. This is sometimes referred to as *jus tertii*, meaning right of a third party. If the defendant could assert the right of the third party as a defense against the plaintiff, it might obviate the necessity for the third party to bring an action. But, even assuming plaintiff is not a holder in due course, the general rule is that the defendant is not entitled to assert the third party's claim as a defense. Exceptions are recognized, however, when (1) the defense is that the plaintiff or someone through whom he holds the instrument acquired it by theft, (2) the payment or satisfaction of plaintiff would be inconsistent with the terms of a restrictive indorsement, or (3) the third party himself defends the action for the defendant. 3–306(d).

Example 63

M delivers his note to P. H fraudulently induces P to negotiate the note to him. M refuses to pay H on the ground that H obtained the note from P by fraud.

H sues M. Judgment for H. Although P had a claim for fraud which entitled him to recover the note from H, or to defend on the ground of fraud if H sued him as an indorser, M is not entitled to assert this defense against H even though H clearly is not a holder in due course. It would have been otherwise if P had defended the action for M, or if the note had been indorsed in blank by P and stolen by H, or if P had indorsed the note on condition that H first deliver some stock to him (one kind of restrictive indorsement) and the stock had not been delivered, for in this case payment would have been inconsistent with the terms of the restrictive indorsement.

§ 3. Claims Asserted Against Holder in Due Course

When an instrument is acquired by a holder in due course, any previous claim against it is cut off. 3–305(1). This is so even though the holder in due course acquired the instrument from a finder or thief who never became the owner of the instrument and who had the power to transfer ownership only because the instrument was payable to bearer.

Furthermore, the holder in due course is protected against not only the claims that might have been asserted against his transferor or any prior party but also against any claim that the *immediate* transferor of the holder in due course might normally have asserted. For example, if an infant or an adjudicated incompetent negotiates an instrument under such circumstances that the transferee becomes a holder in due course, the latter holds the instrument completely free of any claim of the infant or incompetent. This is so even though infancy or an adjudication of incompetency are *defenses* which are good even against a holder in due course. See, e.g., Snyder v. Town Hill Motors, Inc., 193 Pa.Super. 578, 165 A.2d 293 (1960). Similarly, if the holder in due course acquires an instrument in a transaction based on a material mutual mistake, the transferor normally could assert the mistake as a defense if sued by the holder in due course, but could not recover the instrument on the ground of mistake.

The only *claims* that can be asserted effectively against a holder in due course are those which arise after he becomes a holder. For example, if a holder in due course contracts to sell an instrument, but fails to deliver it, the buyer has a claim which he can effectively assert against the holder in due course.

§ 4. Claims Asserted Against Person Lacking Rights of Holder in Due Course

Unless the party having a claim to an instrument loses his claim because of delay or other inequitable conduct or has sacrificed his claim in some other way, he is entitled to recover the instrument from anyone who acquires it until it reaches the hands of a holder in due course. 3–306(a), 3–305(1).

§ 5. Procedure

In the law of commercial paper, especially the area concerned with defenses and claims, procedure bears a close relationship to substantive law and merits at least brief consideration.

Ch. 11   *DEFENSES AND CLAIMS*

A. PLAINTIFF

The proper party to sue on a negotiable instrument normally is the holder who is usually the owner. His status as holder usually is sufficiently established by proof that he is in possession of an instrument that runs to him.

The holder usually is the owner or is acting for the owner. If an instrument is lost, destroyed, or stolen, however, the owner lacks possession and is not the holder. As owner he alone usually is entitled to payment. Recognizing this, the Code permits one claiming to be the owner under such circumstances to recover from anyone who is liable on the instrument, with necessary safeguards.

Since he does not have possession of the instrument, however, he does not enjoy the usual procedural advantages accorded a holder. To recover it is necessary for him to allege and prove that he is the owner, the terms of the instrument, the facts that prevent the production of the instrument and facts establishing defendant's liability. 3–804.

Also, to protect the defendant who is required to pay without actually receiving the instrument against the risk that he may later have to pay a holder in due course or someone else having rights superior to the plaintiff's, the court may in its discretion require the plaintiff to furnish security sufficient to indemnify the defendant against loss that results from his being required to pay a second time. 3–804. The likelihood of the court requiring security depends on a variety of factors such as the length of time elapsed if the plaintiff claims the instrument was lost or stolen and the degree of proof if he claims it was destroyed. Comment, 3–804.

B. BURDEN OF PROOF

Normally a plaintiff has the burden of proving the facts on which his right to recover is based and the defendant has the burden of proving any defenses on which he relies to avoid the liability which normally flows from the facts proven by the plaintiff. Generally the burden of proof is the same when someone sues on a negotiable instrument. An exception is recognized with respect to consideration. Ordinarily a person suing on a simple contract must allege and prove consideration. In an action on a negotiable instrument, lack of consideration is a matter of defense. 3–408. If the defendant contends that he received no consideration, he has the burden of proving this as an affirmative defense. Nicholas v. Zimmerman, 159 Ind. App. 525, 307 N.E.2d 900 (1974). Of course, lack or failure of consideration is not a good defense against a holder in due course.

C. SIGNATURES

Any one who sues on a negotiable instrument relies on one or more signatures. For example, a holder suing a maker relies on the maker's signature and any intervening indorsements necessary to establish his status as holder. A defendant is held to have admitted any signature which he does not expressly deny in his pleading. 3–307(1). The defendant is required to

Ch. 11   *DEFENSES AND CLAIMS*

deny the signature in his pleading if he wants to raise the issue so that the plaintiff is assured of a fair warning and has an opportuny before the trial itself to gather evidence that the signature is genuine. Also, not requiring proof unless the signature is specifically denied in defendant's pleading saves much time and expense; for signatures on negotiable instruments are almost always genuine. Nonetheless, signatures sometimes are denied and plaintiffs then have the burden of proving them. 3–307(1)(a).

Even assuming that a signature is denied, however, the person relying on it has the advantage of a rebuttable presumption which requires the trier of the facts to find that the signature is genuine or authorized unless evidence is introduced to support a contrary finding. 1–201(31), 3–307(1)(b). The presumption arises not only because forged or unauthorized signatures are uncommon but also because evidence relevant to the issue is more likely to be accessible to the person denying the signature. Because this second basis would not apply if a purported signer has died or has become incompetent, the presumption is not recognized in such cases. 3–307(1)(b).

D. ESTABLISHING DEFENSES

When the signatures relied on by the plaintiff have been admitted or established, the mere production of the instrument normally entitles the holder to recover on it unless the defendant sustains the burden of establishing all of the essential elements of a valid defense. 3–307(2). In Telpner v. Hogan, 17 Ill.App.3d

152, 308 N.E.2d 7 (1974), it was held that the burden of proving the defense of payment was not sustained by proving that defendant had delivered to an attorney checks in the amount of notes sued on without proving also that the attorney was duly authorized to receive payment or that the money reached the hands of the holder of the notes.

E. ESTABLISHING STATUS AS HOLDER IN DUE COURSE

If the defendant does not establish a defense, the holder normally is entitled to recover whether or not he is a holder in due course. See, e.g., Blake v. Coates, 292 Ala. 351, 294 So.2d 433 (1974). Once a defense is established, however, anyone asserting the rights of a holder in due course to avoid the effect of the defense has the burden of proving that he, or someone through whom he holds, is in all respects a holder in due course. 3–307(3). See American State Bank v. Richendifer, 36 Or.App. 199, 584 P.2d 323 (1978).

F. THIRD PARTIES

Often a party who loses an action on a negotiable instrument has a right of recourse against some prior party, who in turn may have recourse against a party prior to him, and so on. When a right of recourse exists, it is possible to have the matter settled in several lawsuits. Usually, however, it is more efficient to settle the various issues and determine the various liabilities in a single lawsuit. For the purpose of supple-

Ch. 11   *DEFENSES AND CLAIMS*

menting existing state and federal statutes relating to procedures for interpleader and joinder of parties and to help provide for more efficient disposal of multi-party litigation, Section 3–803 provides:

> Where a defendant is sued for breach of an obligation for which a third person is answerable over under this Article he may give the third person written notice of the litigation and the person notified may then give similar notice to any other person who is answerable over to him under this Article. If the notice states that the person notified may come in and defend and that if the person notified does not do so he will in any action against him by the person giving the notice be bound by any determination of fact common to the two litigations, then unless after reasonable receipt of the notice the person notified does come in and defend he is so bound.

(For some procedural aids relating to proof of presentment, dishonor, notice of dishonor and protest, see Chapter 8, Section 2).

In sum, once our holder has reached the exhalted status of "HDC," the range of defenses to which he is subject is limited to those stated in Section 3–305(2). Even there, reductions in the age of majority and the number of transactions which are "void" as opposed to "voidable" have further limited the risk assumed. Without the protection of HDC status, however, the transferee or holder has the same risk exposure as an assignee, see 3–306, unless Section 3–307 provides some procedural advantages. When all else fails, however, the holder, whether in due course or not, may be able to pass some of the risk back to a prior transferor under the warranty provisions of Section 3–417(2).

CHAPTER 12
DISCHARGE

Chapter 7, entitled "Liability of the Parties on the Instrument," is concerned mainly with the *imposition and nature of the duties* acquired by a person who signs and issues or transfers a negotiable instrument —liability *on* the instrument. The main concern of the present chapter is *how these duties are discharged*—that is, terminated or brought to an end. Before proceeding with discharge of liability *on* the instrument however, it is fitting to consider briefly the circumstances under which the issue or transfer of an instrument might result in the discharge of the *underlying obligation* for which the instrument is issued or transferred.

§ 1. Discharge of Underlying Obligation by Issue or Transfer

It is easy to overlook the fact that an instrument usually is issued or transferred for the purpose of discharging some *underlying obligation*, such as the duty to pay off a loan or to pay for goods, services, or rent. Normally, this underlying obligation is contractual and can be discharged like any other contractual obligation. Thus, it might be discharged by a breach of contract, rescission, release, merger, accord and satisfaction, novation, judgment, intervening impossibility of performance, change of law, and in several other ways. We are concerned here only with determining the circumstances under which the underlying obligation is discharged by the issue or transfer of a negotiable instrument.

One who issues or transfers an instrument to satisfy an underlying obligation is likely to think that the delivery of the instrument immediately and completely discharges him from further liability on the underlying obligation. There are a few situations wherein this assumption is justified, but usually it is not.

The general rule is that the effect of the issue or transfer of an instrument is merely to *suspend* the right of the creditor to sue on the underlying obligation. 3–802(1)(b). If the instrument is payable *on demand*, the right to sue normally is suspended until it is *presented for payment.* If the instrument is payable at *a definite time*, the right to sue is suspended until the instrument is *due.*

Example 64

S sells and delivers goods to B with no express agreement as to the time of payment. Later, S takes B's 90-day note for the price as payment. A few days later, S changes his mind, tenders a return of the note to B and demands immediate payment. B refuses. S promptly sues for breach of the sales contract. Judgment for B. When S took B's note, the right to sue on the underlying contract was automatically suspended until the note became due.

If an instrument is *paid* or the obligor is discharged for any reason from liability on the instrument the underlying obligation is *discharged*; but if it is *dishonored*, the creditor can bring an action either on the instrument *or* on the underlying obligation. 3–802(1)(b).

Example 65

S takes B's check as payment for goods sold and delivered. The check is duly presented for payment at the drawee bank, but is dishonored. S has a choice. He may sue B on his secondary liability as drawer of the check or he may sue B on the underlying obligation and pursue the remedies available to a seller of goods under Article 2.

There are basically two exceptions to the general rule that the issue or transfer of an instrument merely suspends the right to sue on the underlying obligation. The first exception is recognized when it is found that the obligee expressly or impliedly agreed to take the instrument as absolute payment. 3–802(1)(a). The second exception is recognized when the obligee receives an instrument on which a *bank* is drawer, maker, or acceptor and *there is no recourse on the instrument against the underlying obligor.* 3–802(1)(a). When an obligee receives an instrument in either of these two special circumstances, the liability of the obligor on the underlying obligation is immediately discharged. Perhaps it should be emphasized that neither of these exceptions occurs very often. It is not common for a creditor to take even a negotiable instrument with the understanding that the debtor's underlying obligation is absolutely discharged. And even when a debtor tenders an instrument on which a bank is maker, drawer or acceptor, the creditor usually will insist on the debtor's indorsement, in which case the right to sue on the underlying obligation is merely suspended as if no bank had become liable.

Example 66

1. A owes B $1,000. B demands payment. A states that his only asset is a bearer demand note for $3,000 made by C who is known by A and B to be in financial difficulty. B agrees expressly to take the note without A's indorsement as absolute payment of the debt of $1,000. The note is dishonored and proves to be completely worthless. B cannot recover from A on the note or on the underlying obligation. The result would have been the same even if B had not *expressly* agreed to take the note as absolute payment because such an understanding would have been clearly *implied.*

2. C renders services to D. Without indorsing, D delivers to C in payment a certified check drawn by X payable to C. The check is dishonored because, totally unknown to D, the bank is insolvent. C promptly notifies D of the dishonor. C has no rights against D either on the check or on the underlying obligation. A bank is acceptor and there is no right of recourse on the instrument against D, the underlying obligor, who neither drew nor indorsed the check. The result would have been the same if the facts remained the same except that the instrument involved was a cashier's check, an official check or a bank draft. The result would be otherwise, however, regardless of which of the above instruments were used, if C had induced D to indorse, for in this case there would have been recourse on the instrument against D, the underlying obligor, on his irregular or anomalous indorsement. See In re Tri-Power Electronics, Inc., 27 UCC Rep.Serv. 1071 (D.Utah 1979).

Even though the underlying obligation usually is merely suspended and not discharged by the *issuance or transfer* of a negotiable instrument, it is fundamental that *the underlying obligation is totally dis-*

charged whenever the obligor's obligation on the instrument is discharged. 3–802(1)(b). For this reason, it usually is sufficient to focus attention on the obligor's liability on the instrument, knowing that when this is discharged so also is his liability on the underlying obligation.

Example 67

> T owes L $500 for rent. In payment, T, on May 5, 1982, negotiates to L by an unqualified indorsement, a note for $500 made by X and payable to the order of T on June 1, 1982. On July 1, 1982, L for the first time demands payment from X, but the note is dishonored because X is short of funds. L sues T on the *underlying obligation.* L cannot recover from T. As explained in Chapter 8, L's unjustified failure to make a timely presentment when the note became due discharged T, as indorser, from liability on the note even though T cannot show that he suffered any loss as the result of the delay. When T's secondary liability on the note was discharged so was his liability on the underlying obligation.

The obligations that might be incurred on the basis of warranties were discussed in Chapter 9. It will be recalled that even though the secondary liability of an unqualified indorser on the instrument might be discharged by a failure to make proper presentment or to give him due notice of dishonor, he might still be liable on the basis of one or more of the transferors' warranties. This is so even though the underlying obligation of an unqualified indorser was discharged by the discharge of his liability on the instrument.

Example 68

B owes S $5,000. B delivers S in payment a cashier's check for $5,000 on which the payee's name has been indorsed in blank by a forger who sold it to B. B did not indorse. The bank refuses payment because of the forgery of the payee's signature. S sues B. Since B did not indorse the check, he is not liable *on* it. Neither is he liable on the underlying obligation because a bank is drawer and there is no recourse on the check against B. However, as explained in Chapter 9, a transferor of an instrument for consideration warrants that he has title to it. B is liable for breach of this warranty because the instrument is still owned by the original payee whose signature was forged.

The foregoing discussion relates to the question of the time of the discharge of the *underlying obligation* for which an instrument is issued or transferred. The remainder of the Chapter relates to the discharge from liability *on the instrument itself.*

§ 2. Discharge of Liability on Instrument—Basic Propositions

To facilitate an understanding of the law governing the discharge from liability on the instrument, one should recognize five general propositions that pervade the subject and are often sources of confusion. The first is that the "instrument," itself, is not discharged. Rather it is the *liability* of one or more persons on the instrument that is discharged. This is commonly expressed by stating that one or more *persons* have been discharged. Second, the liability of various parties may be discharged at different times. Third, a party's liability may be discharged with respect to one or more persons while remaining with respect to one or more others. Fourth, discharge of liability on an instrument is not always final; liability that has been discharged, can sometimes be revived. Finally, the discharge of any party may not be effective against a subsequent holder in due course "unless he has notice * * * when he takes the instrument." 3–602.

§ 3. Discharge as a Defense

In discussing defenses, it was explained that even a holder in due course is subject to the defense of discharge if (a) he knew of it when he acquired the instrument, (b) it was a discharge in an insolvency proceeding, or (c) it arose at the time that he acquired the instrument or later. It was also explained that one who lacks the rights of a holder in due course is subject to the defense of discharge regardless of its nature, his knowledge of it, or the time that it arose. (See Chapter 11.)

§ 4. Bases for Discharge of Liability on Instrument —In General

Putting to one side the limited exemptions that are given to a holder in due course, a person who is sued on a negotiable instrument may assert effectively a defense of discharge that is based on (1) any agreement or act that would discharge a simple contract for the payment of money, 3–601(2), or (2) any provision of Article 3. Some of these latter bases of discharge have been discussed sufficiently already. For example, it has been seen that a fraudulent and material alteration by a holder discharges the liability of all parties whose contracts on the instrument are changed by the alteration (see Chapter 7, Section 2G); that an acceptance which changes the terms of a draft with the consent of the holder discharges the liability of the drawer and indorsers (see page 158); and that an unexcused delay or failure in making a necessary presentment or in giving a required notice of dishonor discharges the secondary liability of unqualified indorsers (see pages 205–207). There remain to be considered in this Chapter some other bases for discharge provided by Article 3 which have been mentioned already but require further discussion and some bases which have not yet been mentioned.

§ 5. Payment or Satisfaction

By far the most common way of obtaining a discharge from liability on an instrument is by payment or satisfaction. As was mentioned earlier, however, it is fundamental that one can obtain a discharge from liability in this way only by rendering payment or satisfaction to the *proper person*. Normally the proper person is the *holder*. A holder, it will be recalled, is any person who is in possession of an instrument that runs to him. Sometimes a proper person to pay is not, strictly speaking, the holder but someone who legally stands in his position. For example, such person might be an agent, or other representative of the holder or his estate, or it might be a transferee from the holder who, under the Shelter Provision of Section 3–201(1) has acquired the holder's rights.

Example 69

1. X has possession of a note on which the indorsement of P, the payee, is forged. At maturity X obtains payment from M, the maker. M is not discharged from his liability as maker because *X was not the holder*. M is liable to P for conversion. 3–419(1)(c). M's remedy is to proceed against X and prior transferors, including the forger, on the presenters' warranty of title.

2. H is holder of a time note. At maturity H obtains payment from M, the maker, *but does not deliver the note to M*. Subsequently, H negotiates the note to X who has notice that the note is overdue but has no notice that it has been paid. X is not entitled to obtain payment from M. When M paid H his liability was discharged because he *paid the holder* even though he did not obtain possession of the instrument. The discharge can be asserted against X even though he is a holder and had no notice of the discharge. The defense of discharge would not have been valid if M had paid H before the note was due and X had been a holder in due course without notice of the discharge. 3–602.

3. H is holder of a note made by M. H negotiates this note to X for value after maturity. H then persuades M to pay him the amount of the note by fraudulently promising to deliver the note to him later. Although X is not a holder in due course, he is entitled to recover from M. X acquired whatever rights H, as holder, had at the time of the transfer. 3–201(1). The defense of discharge by payment is not valid because M *did not pay the holder and so remained liable as maker*.

A. PAYMENT TO OWNER WHO IS NOT HOLDER

Since a holder is a person who has possession of an instrument that runs to him, an owner might not be a holder because (1) although he has possession, the instrument does not run to him or (2) although the instrument runs to him, he lacks possession, or (3) the instrument does not run to him and he lacks possession. In none of these situations would paying the *owner*, of itself, discharge the payor's liability *on* the instrument.

In situation (1) where the owner non-holder has possession, if the payor pays but does not acquire possession of the instrument and it does not show on its face that it is overdue, the payor runs the risk that it might later be acquired by a holder in due course against whom the payment is not a good defense. But payment to the owner non-holder creates no such risk to the payor if the latter takes possession of the instrument and cancels it.

In situations (2) and (3), wherein the owner lacks possession, the risk of paying a non-holder owner cannot be so easily avoided, because of the possibility that the instrument might already be in the hands of a holder in due course. To enable the owner of the instrument who is out of possession to obtain payment without subjecting him to unreasonable inconvenience and, at the same time, to enable the obligor to make the payment without subjecting him to an unreasonable risk that he might have to pay a second time, Section 3–804 provides:

Ch. 12 *DISCHARGE*

The owner of an instrument which is lost, whether by destruction, theft or otherwise, may maintain an action in his own name and recover from any party liable thereon upon due proof of his ownership, the facts which prevent his production of the instrument and its terms. The court may require security indemnifying the defendant against loss by reason of further claims on the instrument.

B. WHEN PAYMENT TO HOLDER DOES NOT EFFECT DISCHARGE

It is the general rule that payment to the holder discharges the liability of the payor and of any subsequent parties even though the payor knows that someone other than the holder asserts a superior right to the instrument. There are, however, four exceptions to the general rule. 3–603(1). The first exception is recognized when a non-holder, claiming to be owner, furnishes adequate *indemnity* against the risk that a holder will later sue the payor and require him to pay a second time. The second exception occurs when payment or satisfaction to the holder is *enjoined* by a court having jurisdiction over the parties. The third exception occurs when someone other than an intermediary or payor bank pays a holder in a manner inconsistent with the terms of a *restrictive indorsement.* The final exception occurs when a party *in bad faith* pays a holder who acquired the instrument by *theft*, or, unless he has the rights of a holder in due course, holds through someone who acquired the instrument by theft. Usually, the fact that a payor knows of the theft is sufficient to establish bad faith unless he has

[*316*]

good reason to believe that the party being paid has the rights of a holder in due course.

Example 70

1. P holds M's note. By fraud, D induces P to negotiate the note to him. Before maturity, P notifies M of the fraud and orders him not to pay D. Nonetheless, at maturity M pays D. M is discharged. He paid the holder. It would have been otherwise if, instead of merely ordering M not to pay, P had obtained and served a court order forbidding M to pay D or if P had supplied indemnity deemed adequate by M.

2. P holds M's note. Planning to negotiate it, P indorses in blank. T steals the note. Unaware of the theft, M pays T. M is discharged because he paid the holder in good faith without knowledge of the theft. It would have been otherwise if, before M paid T, P had advised M of the theft and requested him not to pay.

3. P holds M's demand note payable to bearer. It is stolen by T who negotiates it to H, a holder in due course. P advises M of the theft and demands payment. Nonetheless M pays H. Although M knew of the theft, his liability was discharged when he paid H because H was a holder in due course and was entitled to payment regardless of the theft.

4. P holds a draft accepted by A. P indorses "Pay H for collection. P," and delivers the draft to H. When H demands payment, A pays the draft by satisfying an obligation owed to him by H. A remains liable. Although A paid the holder, the value he gave was not consistent with P's restrictive indorsement.

5. H holds M's demand note. M pays H but does not obtain the note. Later H negotiates the note to HDC. M is liable to HDC. M's obligation was discharged when he paid the holder, but it was revived

when the note later was acquired by a holder in due course.

§ 6. Tender of Payment

If the obligor on a negotiable instrument makes a proper tender of payment in full to the holder at or after maturity, but the holder improperly refuses to accept the payment, the obligation of the party making the tender is *not* discharged. His obligation continues, but the holder's refusal of a proper tender completely discharges any party who has a right of recourse against the party making the tender. 3–604(2). Furthermore, the party making the tender is discharged to the extent of all *subsequent* liability for interest, costs, and attorney's fees as long as he keeps his tender open. 3–604(1).

The underlying reason for discharging a person who has a right of recourse against the person whose tender is refused is that if the tender had been accepted, the party having the right of recourse would have been completely discharged from liability. It would be unjust to weaken his position because the holder chooses to refuse payment.

Example 71

P negotiates a note to H. At maturity M tenders payment but H refuses it. M's liability is discharged to the extent of future interest, costs and attorney's fees so long as he keeps his tender open. P is completely discharged. However slight the inconvenience or loss caused by requiring P to pursue M, it would have been avoided if H had taken the payment. It is

DISCHARGE Ch. 12

fair that H rather than P suffer this inconvenience or loss.

An instrument frequently is made or accepted to be payable at a particular place—typically the place of business of the maker or acceptor or a bank. *Unless the instrument is payable on demand*, the fact that the instrument's maker or acceptor is able and ready to pay at the place of payment specified when the instrument falls due is equivalent to tender. 3–604(3).

§ 7. Reacquisition

When an instrument is returned to or reacquired by one who previously had been a party to the instrument, any *intervening party* is discharged from liability as against the reacquiring party and any subsequent parties who are not holders in due course without notice of the discharge. 3–208.

Example 72

By successive unqualified indorsements, P, the payee of a draft negotiates it to A who negotiates it to B who negotiates it back to P. A and B are automatically discharged from liability. The reason is that if P were to sue either A or B on that person's secondary liability, the party sued would in turn have a right to sue P. The result would be two unnecessary lawsuits. If, however, after reacquiring the draft, P negotiates it to H, a holder in due course, and the instrument is dishonored, H normally may recover from P, A or B on their secondary liability. The result would be otherwise if, before negotiating the draft to H, P strikes out the indorsements of A and B. In that case H is charged with notice of the discharge of A and B. 3–208 and Comment. Even though notice of the discharge would not prevent H from being a holder in due course, it does subject him to the defense of discharge.

If the instrument is reacquired by one who has himself no right of action or recourse on the instrument against anyone, the liability of all parties is discharged. 3–601(3)(a). The instrument is not, however, completely sterile. It may still be reissued or renegotiated so as to create new liability; and if it is acquired by a holder in due course without notice of a prior discharge, he may enforce it without regard to the prior discharge. 3–602.

Example 73

M signs a demand note for the accommodation of P. The note is promptly negotiated from P to A to B to H by successive unqualified indorsements. H promptly obtains payment from P. The liability of M, P, A and B is discharged. P has no right against A and B, later indorsers; and he has no right against M who signed for his accommodation. But if P, having paid H, promptly negotiates the note to HDC who does not have notice of the prior discharges, the liability of M, P, A and B on the instrument is revived. No discharge of any party under Article 3 is effective against a holder in due course unless he has notice of it when he acquires the instrument.

§ 8. Discharge of Party with No Right of Action or Recourse

In the preceding paragraph it was stated that if an instrument is *reacquired* by a party who has no right of action or recourse on the instrument against anyone, the liability of all parties is discharged. Reacquisition is, of course, one way in which such person might obtain a discharge. The same result normally follows, however, if he is discharged in any other way provided by Article 3. 3–601(3)(b).

Example 74

P, payee of a note made by M, negotiates it to A by unqualified indorsement. A negotiates the note to H by unqualified indorsement. H obtains payment from M whose liability is discharged pursuant to Section 3–603(1). The secondary liability of P and A is automatically discharged. The liability of P and A would not have been discharged if M's liability had been discharged by bankruptcy pursuant to Federal statute rather than by payment or some other act or event pursuant to Article 3.

An exception is recognized when a holder releases a party who has no right of recourse but expressly reserves his rights against a party who has rights against the party being released. 3–606(2).

§ 9. Impairment of Right of Recourse or Collateral

If A owes B $1,000 and C binds himself to pay the $1,000 if A does not, there is a simple suretyship relationship in which A is the principal debtor, B is the creditor, and C the surety.

A well established principle of suretyship law is that if the creditor releases the principal debtor the creditor is treated as if he intended also to release the surety. This intention is not attributed to him if, when he releases the principal debtor, he expressly reserves his rights against the surety; therefore the surety is not discharged. And, so far as the surety is concerned, neither is the principal debtor.

The same general principles are applied by the Code to the more involved relationships that exist between and among the holder and the various parties who incur liability on a negotiable instrument. In a simple case involving only a maker, an unqualified indorser and a holder, the maker is viewed as the principal debtor, the holder as the creditor, and the unqualified indorser as the surety. The case becomes more complicated where there are a number of secondary parties because each indorser has a right of recourse with respect to each prior unqualified indorser; and among them a later party is a creditor, any unqualified indorser who precedes him is a surety, and the primary party is deemed to be the principal debtor.

The Code applies a general principle of suretyship law to the holder and to the various parties who are liable on a negotiable instrument by providing that, "the holder discharges any party to the instrument to the extent that without such party's consent the holder * * * releases or agrees not to sue any person against whom the party has to the knowledge of the holder a right of recourse." 3–606(1)(a).

Example 75

H holds a note made by M and indorsed without qualification by A, B and C in that order. If H releases C, the liability of the other parties is unchanged because none of them has a right of recourse against C. If H releases B, however, C is discharged because C has a right of recourse against B; but A is not discharged because A has no right of recourse against B. If H releases M, then A, B and C are discharged because each of them has a right of recourse against M.

DISCHARGE

The Code provides also that if a holder wishes to do so, he can release a party against whom, to the knowledge of the holder, another has a right of recourse, and at the same time expressly reserve his rights against the person having the right of recourse. 3–606(2)(a). If the holder does so, the reservation of rights is effective; but the party against whom the right of recourse is preserved also retains his right against the party being released. 3–606(2)(a), (c).

Example 76

> P negotiates a note to A by unqualified indorsement. A negotiates the note to H by unqualified indorsement. H releases M, the maker, but expressly reserves his rights against A. A is not released and insofar as A is concerned, neither is P or M. If A pays H, A has a right of recourse against P or M. If A recovers from P, P has a right of recourse against M. The net effect of H's release of M is the same as if he had made an express covenant not to sue M.

The same principles that apply to the complete *release* of a party against whom a person has recourse apply also to a binding agreement by the creditor merely to *extend the time* for the payment of the debt. Wilmington Trust Co. v. Gesullo, 430 A.2d 1084 (Del. Super.1980). If, however, the makers or indorsers agree to an extension by a clause in the note, a subsequent unilateral extension by the holder will not discharge. Whether the surety has in fact consented and the scope of that consent are key issues in most cases. See Union Constr. Co., Inc. v. Beneficial Standard Mortg. Investors, 125 Ariz.App. 433, 610 P.2d 67 (1980)

Ch. 12　　　　　*DISCHARGE*

(indorsers agree that note might be "extended from time to time").

Also in accordance with broad principles of suretyship law, the Code provides that any party to an instrument is discharged to the extent that without his consent the holder "unjustifiably impairs any collateral for the instrument given by or on behalf of the party or any person against whom he has a right of recourse." 3–606(1)(b).

Example 77

1. M borrows $10,000 from P and delivers to P his interest bearing note for that amount and bonds worth $15,000 as security. By an unqualified indorsement, P negotiates the note to H for $10,000 and also delivers the bonds to H. M asks H to return the bonds. Feeling that he can rely on P's secondary liability, H complies. M meets with market reverses and dishonors the note. H gives P prompt notice of dishonor. P is not liable to H. If H had retained the bonds, he could have foreclosed on them or proceeded against P on his secondary liability. If H had required P to pay, P would have been entitled to foreclose on the bonds. By returning the bonds to M, H impaired the collateral to which P was entitled, and therefore P is discharged.

2. M purchased goods from P on credit and issued to P a note for the purchase price of $10,000. P created a purchase money security interest in the goods under Article 9 but did not perfect by filing. P promptly negotiated the note and assigned the unperfected security interest to H. H failed to perfect the security interest by filing. M defaulted and, shortly thereafter, declared bankruptcy. The reasonable value of the goods was $8,000 and C, holder of a perfected security interest with priority over H, satisfied his claim of

$5,000 from the proceeds of sale. Unless P has consented, H's failure to perfect is an impairment of collateral. P is protected to the extent that H's failure actually impaired the collateral. The burden is on P to establish that loss. The loss in this case was $5,000 and, rather than a complete discharge, P's liability as an indorser is reduced by $5,000. Farmer's State Bank of Oakley v. Cooper, 277 Kan. 547, 608 P.2d 929 (1980).

§ 10. Cancellation

The holder of an instrument may discharge the obligation of any party on it by cancelling it in any manner apparent on the face of the instrument. Cancellation may be effected "by intentionally cancelling the instrument or the party's signature by destruction or mutilation, or by striking out the party's signature." 3–605(1)(a). Instruments are cancelled most frequently by marking them "Paid." Although striking out the signature of a party to the instrument discharges the person's liability, it does not affect the holder's title to the instrument even though that signature was essential to his chain of title. 3–605(2).

To be effective a cancellation must be (1) by the holder and (2) intentional. In Liesemer v. Burg, 106 Mich. 124, 63 N.W. 999 (1895) it was held that there was no effective cancellation when the *maker* of a note, who had requested possession to compute the amount due, marked the note "paid" without making full payment.

§ 11. Renunciation

The holder of a negotiable instrument also may discharge any party by renunciation, which may occur in either of two ways: (1) by a separate writing signed and delivered by the holder or (2) by the surrender of the instrument to the party to be discharged. 3–605(1)(b). Since a renunciation does not appear on the face of an instrument, it is possible that a holder in due course will not be bound by it because he is unaware of it when he takes the instrument. Neither cancellation nor renunciation requires consideration. 3–605(1). Consider this case.

> *Example 78*
>
> H holds M's note to which M has no defense. Shortly before maturity, H telephones M and states that he releases him from his liability on the note and that he considers the note to be void. Later that day H has a change of heart and sells the note to A. At maturity, M refuses to pay. A sues. Defense: Renunciation by H. Judgment for A. The defense of renunciation requires that the holder either deliver to the obligor a signed writing containing the renunciation or surrender the instrument. It makes no difference whether or not A was a holder in due course, since M had no defense. In fact, M would have had no defense even if H had retained the note and sued. Even if the renunciation had been in writing, it would not have been binding on A if A could have established that he was a holder in due course without notice of it.

Suppose that the holder of a note writes the maker, "I will surrender my notes in the amount of $8500 to (the maker). (signed) Holder." Was this an effective renunciation? In Gorham v. John F. Kennedy College,

Inc., 191 Neb. 790, 217 N.W.2d 919 (1974), it was held that it was not. It was merely a promise and not an outright renunciation so it was ineffective even though in writing.

CHAPTER 13

CHECKS, CHECK COLLECTION AND RELATIONSHIP BETWEEN BANKS AND THEIR CUSTOMERS— HEREIN OF ARTICLE 4

A check is a draft drawn on a bank and payable on demand. 3–104(2)(b). In general it is governed by the same principles applicable to other drafts. There are, however, a number of important problems that arise because of the relationship between Drawer and Payee and their Bank(s). These problems are governed by Article 4 of the Code. 4–102(1). Drawer will have a contract of deposit with Drawee Bank, called a Payor Bank in Section 4–105(b), that obligates Payor Bank to pay checks presented by Payee or his order if there are sufficient funds. See 4–401. Payee will have an account where his bank agrees to make reasonable efforts to obtain the payment of checks deposited for collection. This bank is called a Depositary Bank when it is the first bank to which a check (or "item" see 4–104(1)(g)) is deposited for collection, 4–105(a), and a Collecting Bank when it handles the item for collection. 4–105(d). Both Drawer and Payee are called Customers. 4–104(1)(e). We will use this terminology in the discussion which follows.

Despite the complexity of these problems, it is important to understand the fundamentals of Article 4 and its mix with Article 3. We will start with a brief overview of the check collection process. Next we will examine the relationship between the Payee or his order who deposits an item for collection and the different banks in the collection process. Finally, we will examine the relationship between Drawer and Payor, with particular emphasis upon transactions where something goes wrong. For example, if Drawer issues a check to Payee which is stolen and altered by Thief and Thief obtains payment from Depositary Bank which, in turn, obtains payment from Payor Bank, what is the liability of Payor to Drawer? What rights, if any, does Payee have against Payor, Drawer or Thief? Confronting these questions from Payor's perspective will help to understand the relationship between Articles 3 and 4 and to review many of the basic principles discussed earlier in the book.

§ 1. Overview of the Check Collection Process

Suppose that Drawer issues a check for $5,000 to the order of Payee drawn on Payor Bank as payment for goods sold. How will this check be paid?

If Payor is a local bank, Payee may collect it himself. Within a reasonable time after issue, Payee may visit Payor during business hours, present the check and demand payment. See 3–503(2) & 3–504(2)(c). Payor may require Payee to exhibit the check and identify himself. 3–505(1). Payor may also take time to determine whether the check is properly payable from Drawer's account. In theory, Payor can defer payment without dishonor up to midnight of the banking day of receipt. 4–302(a). In practice, the status of Drawer's account can quickly be determined. If the

Ch. 13 CHECKS AND CHECK COLLECTION

check is in order and there are sufficient funds, the check will be paid ("settled" in Article 4 parlance) in cash "over the counter." This payment is final, 4–213(1)(a), unless Payee has breached a warranty of presentment. Since Payee is not a Customer in this case, see 4–104(1)(e), the presentment warranties of Section 3–417(1) rather than Section 4–207(1) apply. See Chapter 9, Section 2. Compare 3–418.

More probably, Payee will use the services of a bank for collection rather than collect it "over the counter." Payee will deliver the check with a restrictive indorsement, such as "for deposit only" or "for collection," to Depositary Bank, with which he has an account. Depositary Bank will, as a Collecting Bank, see 4–105(d), act as Payee's agent for collection. If Collecting Bank is located in the same city or region as Payor Bank, the check will be presented for payment through a local or regional clearing house. If the banks are some distance apart, Collecting Bank will use intermediary banks, see 4–105(c) and the collection services of the Federal Reserve System, with its detailed regulations and operating letters. In this process, Federal Reserve regulations and operating letters and various clearing house rules have the effect of agreements among the parties, whether assented to by all or not. 4–103(2). Compare 4–103(1).

Upon presentment, Payor Bank has until midnight of the next banking day following the banking day of receipt to pay or dishonor the item. This is the so-called "midnight deadline." 4–104(1)(h). During this period, Payor will, with the aid of a magnetic ink character recognition system (MICR), see Chapter 14, Section 2A, and a computer, verify signatures, ascertain that sufficient funds are available and determine whether the item is otherwise payable. The practice of sorting and proving items on the day received and if a decision to pay is made, posting them to Drawer's account on the second day is called "deferred posting." Comment 1, 4–301(1). To avoid accountability for the item, however, Payor Bank must make a provisional settlement for it before midnight *on the day that it is received* and if a decision to dishonor is made, either return the item or send a notice of dishonor before its "midnight deadline," i. e., midnight of the banking day following receipt. 4–302(a). A proper dishonor revokes the provisional settlement. (More on this later.). A decision to pay is manifested in several ways. See 4–213(1). In check collection, an item is finally paid by Payor Bank when it has "completed the process of posting the item to the indicated account of the drawer." 4–213(1)(c). See 4–109. Upon final payment, Payor Bank becomes accountable for the amount of the item, subject to any breach of presentment warranty by Collecting Bank or its Customer. 4–207(1). Payment in these cases is not made by the physical transfer of currency. Rather, credit balances maintained by member banks and others in federal reserve banks are transferred by check or wire or through an automated clearing house. See Chapter 14, Section 3A. For a more detailed discussion, see Speidel, Summers and White, Commercial and Consumer Law 1369–1397 (3d ed. 1981).

Ch. 13 *CHECKS AND CHECK COLLECTION*

§ 2. Rights of Customer in Collection Process

What are the legal consequences of Payee's decision to use the collection services of Depositary Bank and, say, the Federal Reserve System? Let us examine a few of them in a routine transaction. The focus will be on the Payee's rights as a Customer in the collection process.

A. COLLECTING BANK AS AGENT FOR COLLECTION

Upon delivery by Customer of a check with a restrictive indorsement, "for collection only," Collecting Bank becomes a holder of the item. See 3–205(c), 3–206(1), 4–201(2) and 4–205. But Customer is still the owner of the item. Collecting Bank is an agent of Customer, 4–201(1) with restrictions upon what can be done with the item and how value received is to be applied. 3–206(3).

B. DUTIES OF COLLECTING BANK

As an agent for collection, Collecting Bank has some very specific duties to perform. First, Collecting Bank has a general duty of ordinary care in presenting an item to Payor Bank, sending a notice of dishonor or making an appropriate settlement upon final payment by Payor Bank. 4–202(1). This duty cannot be disclaimed by agreement, although the parties can agree to standards of responsibility that are not manifestly unreasonable. 4–103(1). The measure of damages for the failure to exercise ordinary care is "the amount of the item reduced by an amount which could not have been realized by the use of ordinary care, and where there is bad faith * * * other damages, if any, suffered by the party as a proximate result." 4–103(5). Second, Collecting Bank must send the item by a reasonably prompt method, taking into account a number of relevant factors, e. g., instructions, nature of the item, cost of collection and methods in general use. 4–204(1). The Collecting Bank may send the item directly to Payor Bank or, if authorized, use appropriate clearing house facilities. 4–204(2). Third, if Payor Bank makes final payment, Collecting Bank becomes accountable to Customer "for the amount of the item." 4–213(3). In this case, Collecting Bank must use ordinary care in settling with Customer for the item. 4–202(1)(c). Fourth, if Payor Bank dishonors the item it must exercise ordinary care in sending a notice of dishonor to Customer. 4–202(1)(b). Note that if Collecting Bank fails to act seasonably in sending notice, see 4–202(2), it is liable to Payee for damages rather than the amount of the item.

C. PROVISIONAL SETTLEMENT

A provisional settlement is a payment, credit or remittance that is conditional upon final payment of the item by Payor Bank. Collecting Bank may permit Customer to draw against an item deposited for collection before it is finally paid. Suppose that Customer draws $2,500 against a $5,000 item. At this point, Collecting Bank has given value and, all things being equal, could be a holder in due course of the item. 4–208 and 4–209. See Chapter 10, Section 13. If the item is paid by Payor Bank, Collecting Bank becomes accountable to Customer for the amount of the item and the provisional settlement becomes final. 4–213(3). If the item is dishonored by Payor Bank, Collecting Bank may revoke the provisional settlement and "charge back" the amount of any credit given Customer in the process of collection. 4–212(1). To protect against risks, Collecting Bank has a security interest in the item and will probably have the status of a holder in due course. 4–208.

The situation with Payor Bank is different. Assume that the $5,000 item is presented through the Federal Reserve System for payment at 10 AM on July 5. There is no demand by Payee for immediate payment "over the counter." See pages 329–330. Payor Bank *must* make an appropriate settlement, see 4–211(1) before midnight of the day of receipt, July 5. If this settlement is not made, Payor Bank is accountable for the item. 4–302(a). But, note carefully, this settlement is provisional. Payor Bank has until midnight of the next banking day after receipt, July 6, to make a final decision to pay or dishonor. This is Payor Bank's so-called "midnight deadline." 4–104(1)(h). If Payor Bank fails to dishonor by the "midnight deadline" it has made final payment. 4–213(1)(d). If a dishonor is timely, however, Payor Bank may revoke the provisional settlement and recover any payments made. 4–301(1). The "midnight deadline" is designed to give Payor Bank a reasonable time to process and evaluate the item and to permit the various computer programs to run their course. Comment 1, 4–301. Since most checks are paid, the provisional settlement on the first banking day compresses somewhat the time between deposit and payment, thereby reducing the "float." If the check is paid, the provisional settlement becomes final and the collection process is completed, subject to any breach of presentment warranties.

D. FINAL PAYMENT

This is somewhat complicated. One side of the coin is Payor Bank's failure to dishonor within the "midnight deadline." Thus, if Payor Bank made a provisional settlement on July 5 and *failed* before its "midnight deadline" either to return the item or send a written notice of dishonor, final payment would be made. 4–301(1) and 4–213(1)(d). Payor Bank would be accountable for the amount of the demand item. 4–213(1). On the other side of the coin, Payor Bank can make final payment by taking affirmative action before the midnight deadline expires. Final payment occurs when the bank has done any of the following, "whichever happens first:" (a) paid the item in cash; (b) settled for the item without reserving or having the right to revoke the settlement; and (c) completed the process of posting the item to Drawer's account. 4–213(1). The first two acts of final payment are fairly simple to understand. The third, completing the process of posting, poses some difficulties. Here is an illustration.

Suppose that Payee's check for $5,000 was presented by Collecting Bank to Payor Bank on July 5 and a provisional settlement was made. The check was processed by the computer on the evening of July 5 and stamped "paid." Drawer's account was debited.

Ch. 13 CHECKS AND CHECK COLLECTION

On the morning of July 6, the bookkeeper examined the computer printout and the check and discovered an error: there were insufficient funds in Drawer's account. Payor Bank then reversed its action and returned the item before the "midnight deadline." In the meantime, Collecting Bank had made a provisional settlement with Customer who now is insolvent. Collecting Bank, now a holder in due course, argues that Payor had completed the process of posting and was accountable for the item. Whether Collecting Bank is correct turns on Section 4–109, which provides:

> The "process of posting" means the usual procedure followed by a payor bank in determining to pay an item and in recording payment including one or more of the following or other steps as determined by the bank:
> (a) verification of any signature;
> (b) ascertaining that sufficient funds are available;
> (c) affixing a "paid" or other stamp;
> (d) entering a charge or entry to a customer's account;
> (e) correcting or reversing an entry or erroneous action with respect to the item.

If Payor Bank's usual procedure is to have a careful review of decisions made in the initial computer run and a basis for dishonor is found in that review, it seems clear that the "process of posting" has not been completed. At the very least, Payor's procedure provided the opportunity to correct or reverse decisions at the earlier stage. The harder question is what is the "usual procedure" and when does it end? Suppose, for example, that a careful review of the computer action and the item found no basis for dishonor. The check was stamped paid, the account debited and the item returned to Drawer's file. Later it was fortuitously discovered that Drawer's signature had been forged. If notice of dishonor could still be sent before the "midnight deadline" has final payment nevertheless been made under 4–213(1)(c)? The answer is no if "usual" internal procedures reserved this third, albeit fortuitous, review. If the procedures are silent or ambiguous, there is some disagreement over the proper outcome. One line of argument, usually made by bankers, is that Payor should have up to the "midnight deadline" to correct or reverse an erroneous decision unless the "usual procedure" is clearly to the contrary. They rely primarily upon the language of Section 4–109(e). But that section simply indicates that "usual procedure" may consist of one or more of the listed steps. It does not say that in the absence of a clear procedure the process of posting shall consist of them all. Thus, others argue that in the absence of clarity in procedure, the process of posting should be completed when the check is stamped paid and returned to the file after the second review. That argument, of course, favors Payee and Collecting Bank over Payor.

The analysis is further complicated by Section 4–303. Suppose that Payor's usual procedure clearly contemplated a review that could extend up to the "midnight deadline." The bookkeeper had reviewed the computer run and the item, found no error, and returned the item stamped "paid" to the file. Payor then discovered that Drawer was in default on an obligation owed to Payor and, before the "midnight dead-

Ch. 13 CHECKS AND CHECK COLLECTION

line," reversed the paid entry, set off the defaulted obligation against the bank account and sent a timely notice of dishonor. As between Payee-Customer and Payor, has final payment been made?

Under Section 4-303(1)(d), the set off comes too late (and Payor is accountable for the item) if Payor has first either completed the process of posting or "otherwise has evidenced by examination of such indicated account and by action its decision to pay." Payor has not completed the "usual procedure" in the posting process. But it has, by review and conduct, indicated a decision to pay. In the view of many, this decision forecloses a subsequent reversal under Section 4-303(1). One basis for this outcome is that the language of 4-303(1)(d) differs from that found in 4-213 and 4-109. A decision to pay can be made before the usual process of posting is completed. Another basis is that 4-303 is a priority section. The contest is between Payee's claim that final payment has been made and four external events, a stop order, set off by Payor, service of legal process or notice of insolvency. Priority is determined by "first in time" and if a decision to pay (without error) is first manifested, giving Payor discretion subsequently to reverse the decision in its own interest undercuts the priority policy. Although there are cases to the contrary, this reasoning finds support among the commentators. See, e. g., White and Summers, Uniform Commercial Code § 17-7 (2d ed. 1980). Contra: West Side Bank v. Marine Nat'l Exchange Bank, 37 Wis.2d 661, 155 N.W.2d 587 (1968).

Remember. Even though a payment may be final under 4-213, Payor may have a possible recovery for

CHECKS AND CHECK COLLECTION Ch. 13

breach of presentment warranty against Collecting Bank or Payee. See 4-207(1), 3-417(1) and 3-418.

Example 79

1. Drawer in Ohio drew a $5,000 check on Payor and issued it to Payee in Indiana. Payee deposited it for collection in Depositary Bank and the item was presented to Payor at 10 AM on August 1 for payment. An initial computer run showed that the check should be paid and, before midnight on August 1, Payor settled for the item without reserving the right to revoke the settlement. On August 21, Payor, as was its usual procedure, reviewed the computer run and discovered that there were insufficient funds in Drawer's account. Payor then reversed the entry and sent a notice of dishonor before midnight on August 2. Payor is accountable to Payee for the item. Even though the process of posting was not complete, Payor made final payment by first settling for the item without reserving the right to revoke. 4-213(1)(b).

2. Suppose, in the example above, that the Payor and Depositary banks were private correspondent banks that did not use the Federal Reserve collection system. They agreed, however, to be bound by Federal Reserve operating circulars. Circular No. 6 stated that items in the amount of $2,500 or more must be dishonored by wire. Drawer's check for $5,000 was presented for payment on August 1 and a provisional settlement was made before midnight of that day. On August 2, before the process of posting was completed, it was determined that Drawer's signature was forged. Payor sent a written notice of dishonor before the "midnight deadline" and Collecting Bank seasonably informed Payee of the dishonor. Payor argues that notice of dishonor was proper and timely under 4-301(1)(b). See 3-508(3) (notice of dishonor may be written or oral). Payee argues that Circular No. 6 of

the Federal Reserve as agreed to by Payor and Depositary Bank is an agreement altering the effect of 4–301(1)(b) and having effect even though not specifically assented to by all parties interested in the item. See 4–103(1) and (2). Payee is correct on the scope of the agreement and its legal effect. Payee, by requesting Depositary Bank to collect the item, was bound by reasonable and usual arrangements made by its agent in the collection process. To avoid final payment and accountability for the item, Payor must revoke a provisional settlement in the "time and manner permitted by statute, clearing house rule or agreement." On these facts, a revocation by written notice rather than by wire was improper. 4–213(1)(d).

3. Suppose, in the example above, that Payor's written procedure for posting reserved the right to review and reverse an entry "for any reason" up to the "midnight deadline." Payee's check was received on August 1 and a provisional settlement was made before midnight of that day. On August 2, the bookkeeper examined the computer run and the check and determined that the check should be paid. It was stamped paid and returned to the file. Later that day, the bookkeeper fortuitously discovered that Drawer's signature was forged. No stop order had been issued. Payor then reversed the entry and sent a proper notice of dishonor before the midnight deadline. On these facts, final payment has not been made. The test of 4–213(1)(c) is whether Payor has completed the process of posting and, under the clear internal procedures, the answer is no. See 4–109. But if Payor had received notice that Drawer was insolvent or a creditor of Drawer had served legal process, a different result is arguably dictated by 4–303(1)(d). Payor, without error in its internal evaluation, has indicated a decision to pay. Even though the process of posting may not be completed, that decision to pay was made before the external notice or process was received. This estab-

lishes priority for Payee and the reversal of the entry comes "too late."

§ 3. Contract Between Payor Bank and Checking Account Customer

The relationship between a bank and its checking account customer is the result of a contract voluntarily entered for the mutual benefit of the parties. Such a contract usually is made in an informal manner with a minimum of express terms defining the rights of the parties. Typically, an application to open an account is made by the customer to the appropriate officer at the bank, the customer's needs and the size of the initial deposit are discussed, identification and references are submitted, signature cards are signed, the deposit is made, and a receipt is issued to the customer. Usually, by signing the signature card, the customer agrees to be bound by the rules of the bank. These rules are usually stated on the deposit receipt or on a separate card delivered to the customer. There is no other express contract setting forth the other terms of the relationship. Generally, these terms are supplied by law and custom. In fact, apart from statements relating to bank charges, even the rules which are expressly set forth normally are merely statements of the prevailing law and custom.

Although the parties have much freedom in determining the terms of their agreement, the Code expressly provides that no bank can effectively disclaim responsibility for its lack of good faith or failure to exercise ordinary care, nor can it limit the measure of damages for such lack or failure. But the parties may

Ch. 13 CHECKS AND CHECK COLLECTION

by agreement determine the standards by which the bank's responsibility is to be measured if such standards are not manifestly unreasonable. 4–103(1).

Subject to this agreement, Payor "may charge" against Drawer's account "any item which is otherwise properly payable from that account even though the charge creates an overdraft." 4–401(1). The negative inference is that Payor may not charge the account if the item is not properly payable. Thus, if Drawer's signature or Payee's indorsement have been forged, the instrument has been altered or a valid stop payment order has been issued, payment by Payor and a charge against Drawer's account is a breach of the deposit contract. See Cincinnati Ins. Co. v. First Nat'l Bank of Akron, 63 Ohio St.2d 220, 407 N.E.2d 519 (1980).

A. CHECKING ACCOUNT BANK DEPOSIT

The deposit made by the checking account customer sets up a debtor-creditor relationship between the bank and the customer. The identity of each deposit is lost; the funds become part of the general assets of the bank and the bank becomes a debtor of the customer. In case the bank becomes insolvent, the customer is only a general creditor although he normally has the advantage of his account being insured by the Federal Deposit Insurance Corporation. Sometimes special accounts in the nature of trust accounts are deposited in banks creating relationships other than debtor-creditor.

B. ISSUANCE OF CHECK IS NOT AN ASSIGNMENT

Normally, if a creditor makes an assignment of money owed to him, the debtor becomes legally bound to pay the assignee as soon as the assignee gives him notice of the assignment. If the debtor refuses to pay the assignee, the latter can sue and obtain a judgment against the debtor. As with other drafts, however, merely issuing a check does not operate as an assignment of funds in the drawee's hands. Consequently, even though a bank has adequate funds on deposit to pay checks, it is not liable to the holder of the check unless it accepts the check by certifying it. 3–409(1). The holder has no recourse against the bank. The bank's only obligation is to its customer if it fails to pay when funds are available. The usual remedy of a holder of a dishonored check is against the drawer and indorsers on their secondary liability. (For some theories on which a drawee might be held liable to a holder in *unusual* circumstances, see pages 153–155)

C. CERTIFICATION OF CHECK

Certification is the acceptance of a check by the drawee bank. 3–411(1). By certification, the bank, as an acceptor, becomes primarily liable. Since certification of a check involves the assumption of a new obligation by the bank, it rests in the bank's discretion. Unless otherwise agreed, the bank has no obligation to either the depositor or a third party to certify. 3–411(2). Nevertheless the practice of certifying checks for *drawers* is rather widespread. Even when

Ch. 13 *CHECKS AND CHECK COLLECTION*

they are willing to cash checks, however, many banks are unwilling to certify them for *holders*. Some banks certify checks for holders only with the drawer's approval. Many banks charge when certifying a check for either a drawer or a holder; some charge only a holder; and some charge neither. Many banks charge for certifying only checks drawn on special accounts.

In most cases wherein the certification is obtained by the *drawer*, the reason is that it is required by the payee who is unwilling to rely on the credit of the drawer alone. Contracts commonly provide that payment must be made by certified check, particularly if the amount is large and the payee cannot afford to take the risk that the check will be dishonored.

Normally, when the *drawer* obtains the certification, the payee obtains both the primary obligation of the bank and the secondary obligation of the drawer who stands behind the certification warranting that the certification is good and agreeing to make good if the bank does not. Comment 1, 3–411(1). When a holder procures certification, the situation is different; the obligation of the bank remains the same but the drawer is relieved of secondary liability. 3–411(1). Also, any indorsers who have indorsed prior to certification are discharged when the holder obtains certification because they are deprived of rights against the drawer who has been discharged. 3–411(1).

Example 80

Drawer drew a check for $5,000 on Payor Bank and issued it to Payee. Payee negotiated the check to Holder who presented it to Payor and demanded acceptance. Payor refused. Holder then demanded payment and Payor refused. Even though Payor's conduct violated the deposit contract with Drawer, Holder has no claim against Payor. Holder may sue Drawer and Payee, the indorser, on the instrument or Drawer on the underlying obligation. Drawer may claim damages caused by Payor's wrongful dishonor. If Payor had accepted or certified the check, however, Payor would be liable on the instrument and Drawer and Payee would be discharged. 3–411(1).

§ 4. Bank's Liability for Wrongful Dishonor

Under the contract between the bank and its customer, the bank has a number of duties, but its primary duty is to honor the depositor's checks when he has sufficient funds on deposit for this purpose.

A bank is liable to its customer for the wrongful dishonor of his check. 4–402. Since liability attaches only if the dishonor is wrongful, it does not attach to a refusal to pay when there are insufficient funds in the drawer's account, when a necessary indorsement is missing, or for any other good reason.

Before the Code the legal theories on which banks were held liable to a drawer whose check was wrongfully dishonored included breach of contract, negligence, wilful wrong, libel, and slander. The damages recoverable have varied depending largely on the theory adopted. The Code does not adopt any single theory, but does distinguish between cases in which dishonor is the result of mistake and other cases. Most wrongful dishonor results from mistake. In such cases prior to the Code, merchants or traders sometimes were allowed to recover substantial damages without proof of actual damages, but others were required to prove actual damages. The Code limits recovery to *actual* damages in all cases involving innocent mistake. It recognizes, however, that actual damages may include damages from injury to credit, damages resulting from arrest or prosecution for issuing bad checks, and other consequential damages. The burden of proof in these cases is on the customer.

If the wrongful dishonor is deliberate rather than mistaken, Customer may have a claim for punitive damages if they were available under state law prior to the Code. Section 4–402 is silent on the question and cannot be said to preempt the matter. Section 1–106(1) provides that punitive damages are available if "specifically provided * * * by other rule of law." Thus, if state law provides for punitive damages in extreme cases of wrongful dishonor, they are available to supplement actual damages proved under Section 4–402. See 1–103.

Example 81

Customer, Yacht Club Sales and Service, drew eight checks against an account in Payor Bank. At all times the account had sufficient funds. Before the checks were presented, Payor received a writ of execution on funds in the account of Yacht Club, Inc., a separate customer. If Payor had paid funds from the account of Yacht Club, Inc. after receiving the writ it would be liable to the judgment creditor. Payor was not sure which account had been attached by the writ and, pending clarification from the court, put a hold on the accounts of both Yacht Club Sales and Service and Yacht Club, Inc. Neither customer was informed of the hold. The checks of Yacht Club Sales and Service were subsequently dishonored upon presentment. In a suit for damages under Section 4–402, it was held that the wrongful dishonor was not mistaken as a matter of law. Payor had deliberately protected its own interest without giving notice to Customer. Thus, Customer could claim both actual damages and, if state law permitted, punitive damages. The burden of proving actual and punitive damages, however, was on Customer. Yacht Club Sales and Serv., Inc. v. First Nat'l Bank of North Idaho, 101 Idaho 852, 623 P.2d 464 (1980).

§ 5. Charging Customer's Account

As a special kind of draft, a check is an order to pay. When a customer issues a check he orders his bank to pay it according to its terms. Therefore, if the check is in proper form and the bank carries out the order in good faith, the bank may properly charge the customer. 4–401.

A. OVERDRAFT

When a bank pays a check issued by its customer in proper form, it properly may charge his account even though it is an overdraft. 4–401(1). Although a customer normally has no right to overdraw his account and the bank is not obliged to pay an overdraft, a check that overdraws an account is nonetheless an order to pay.

The order carries with it an implied authorization to charge the customer's account and an implied promise by the customer to reimburse the bank. Comment 1, 4–401. Allowing the bank to charge the drawer's account becomes especially important when it pays an overdraft to a holder in due course or to one who later

Ch. 13   *CHECKS AND CHECK COLLECTION*

changes his position in good faith in reliance on the payment because in those circumstances there is no breach of any of the presenters' warranties and the payment is final. 3–418 and 3–417(1). If the bank could not charge the drawer's account, it would have no security for the extension of credit. Whether or not the account is charged, Customer has a legal obligation to repay the overdraft.

B.   DRAWER'S SIGNATURE UNAUTHORIZED

Normally, Payor can charge Customer's account only if his signature was authorized. Otherwise he did not give an order to pay. If Payor pays over a forged Drawer's signature, Customer's account, if debited, must be recredited. In most cases, Payor will bear the full risk. If the presenter was the forger, he will in all probability be long gone. If the presenter was an innocent transferee of the check from the forger, he warrants that he has "no knowledge that the signature of the * * * drawer is unauthorized." 4–207(1)(b). He does not warrant that all signatures are genuine. Thus, if the transferee has changed in position in good faith reliance on the payment, the payment cannot be recovered by Payor. 3–418.

C.   PAYMENT TO NON–HOLDER

The drawer's order to pay is qualified. To be effective so as to give the bank the right to charge the depositor's account, the bank's payment normally must be made to the payee or his order. This will be a *holder* or someone authorized by the holder to receive payment. The bank, therefore, must determine who is the holder. If the instrument has not been indorsed, this usually presents no problem so long as the bank is acting in good faith. If such an instrument was issued as bearer paper (typically "pay to bearer" or "pay to the order of cash"), the holder is anyone in possession of the instrument. If the instrument was issued as order paper, the payee is the holder if he has possession. If the instrument has been indorsed, however, the bank must make certain that all indorsements in the "chain of title" to the presenting party are authorized. If any one of these indorsements is not authorized, the presenter is not the holder and the bank cannot properly charge the drawer's account if it pays. The bank's remedy is to proceed on the presenters' warranty of title given by the presenter and all transferors back to and including the wrongdoer, all of whom lacked title. Whether or not the bank collects on the basis of the presenters' warranty of title, it may be held liable for conversion to the party whose indorsement was forged. 3–419(1)(c). But in this case the bank normally will be entitled to charge the drawer's account on the theory that the bank is subrogated to the rights of the party whose indorsement was forged. 4–407. (See subrogation, pages 360–363)

Example 82

Drawer, in payment for goods received, drew a check for $5,000 on Payor Bank and issued it to Payee. Thief, without Payee's negligence, stole the check and forged Payee's indorsement. Thief then transferred the check to Transferee, who took in good faith, without notice and for value. Transferee presented the check and was paid by Payor over the forged indorsement. The payment was wrongful as to Drawer. Transferee, however, was neither the holder nor the owner of the item. Payor, therefore, can recover from Transferee for breach of the warranty of good title, 4–207(1)(a), even though Transferee has relied in good faith on the payment. Transferee can in theory then recover from Thief. If Payor is unable to recover any or all of the loss from Transferee, Payor, to the extent of the loss, may claim subrogation to the rights of Payee under Section 4–407(b). The effect is that Drawer bears the risk of Payor's payment over the forged indorsement, in that Drawer still owes Payee for goods received and Payor can refuse to recredit Drawer's account. In practice, Payor will recredit Drawer's account and resort to insurance covering the risk of forger indorsements.

D. AMOUNT TO BE CHARGED

If the bank pays a proper party on a check which has *not* been altered, it may, of course, charge the depositor's account the full amount of the check.

Instruments which have been altered, however, present a special problem. If an instrument is materially altered by the holder with a fraudulent intent, each party whose contract is affected is discharged from liability except as against a subsequent holder in due course. Accordingly when a check is so altered, the drawer is discharged.

If the check thereafter is transferred to a holder in due course, however, the latter is entitled to recover according to its original tenor if it was complete when issued, or as completed, if it was incomplete when issued.

If the bank acts in *good faith* when making a payment to a holder, the bank is entitled to charge the drawer's account according to the original tenor of the altered check if it was complete when issued, and according to the tenor of the completed instrument if it was incomplete. 4–401(2). The bank is so entitled whether payment was made to a holder in due course who is entitled to the payment or to the defrauder or some other holder who is not.

At first glance, it may seem unfair to the drawer to allow the bank to charge his account when it pays a holder other than a holder in due course after the drawer has been discharged by an alteration. It does not seem unfair, however, when seen through the eyes of the bank which pays in good faith. If the instrument was complete when issued, the drawer really suffers no loss when he is required to pay according to the original tenor of the check; he is merely prevented from getting a windfall at the expense of the bank. If the check is incomplete when issued, requiring the drawer to pay more than he intended seems fairer than thrusting any part of the loss on the bank. The drawer signing an incomplete instrument should realize that he is courting trouble whereas the bank is merely doing what is expected of it—paying a holder, in good faith.

In a case wherein a bank is not entitled to charge a drawer's account for the full amount it has paid on an altered check, it is entitled to recover its loss from the wrongdoer or any other party who broke the presenters' warranty against alteration. (See pages 229–232)

Ch. 13 *CHECKS AND CHECK COLLECTION*

§ 6. Drawer's General Duty of Care

The drawer loses the right to bar the drawee from charging his account for a payment made on an altered instrument in excess of the amount originally provided or on an unauthorized signature of his own or of an indorser in the chain of title if the drawer's own negligence substantially contributed to the alteration or unauthorized signature and the drawee bank paid in good faith and in accordance with reasonable commercial standards of the banking business. 3–406.

§ 7. Customer's Duty to Discover and Report Unauthorized Signature or Alteration

In addition to the general duty of care described above that is imposed on a drawer who wishes to bar the drawee bank from charging his account on the basis of an alteration or unauthorized signature, the Code imposes a more precise duty in handling his bank statements.

Most banks furnish their checking account customers with periodic statements of account accompanied by cancelled checks and other items in support of debit entries. It is to the interest of the customer to check the cancelled checks against his check stubs and verify the balance. If he does so, he is almost certain to detect any alterations made in the checks he has issued and any checks he did not sign or authorize.

If he finds that the bank has paid and charged a check on which his signature was unauthorized, the customer is entitled to have the bank re-credit his account unless he has ratified or is precluded by his negligence from asserting the unauthorized signature. Similarly, if he finds that an item has been altered by being raised, he may be entitled to require the bank to re-credit him for any excess paid over the original tenor of the check. Also, if the bank has paid someone other than the holder, relying on a forged indorsement in the chain of title, the customer is entitled to have the bank credit his account for the amount of the payment.

A customer should act with reasonable promptness in discovering and reporting any such discrepancies. If he fails to do so, his bank may be injured in two ways. First, it may be deprived of the opportunity to intercept the wrongdoer and obtain restitution. Second, through lack of knowledge of the wrongdoing, it may be unable to prevent future losses at the hands of the same wrongdoer.

To protect the bank from incurring losses which can be expected to flow from a customer's careless handling of his bank statements and to induce the customer to exercise care the Code provides in Section 4–406(1):

> When a bank sends to its customer a statement of account accompanied by items paid in good faith in support of the debit entries or holds the statement and items pursuant to a request or instructions of its customer or otherwise in a reasonable manner makes the statement and items available to the customer, the customer must exercise reasonable care and promptness to examine the statement and items to discover his unauthorized signature or any alteration on an item and must notify the bank promptly after discovery thereof.

If the customer fails to exercise reasonable care and promptness in discovering and reporting these discrepancies, he is precluded from asserting against his bank either *his own* unauthorized signature or any alteration on any item on which the bank can show that it suffered loss because of his failure. 4–406(2)(a).

In addition, by failure to perform this duty, the bank's customer is precluded from asserting against the bank any alteration or unauthorized signature—*an indorser's as well as his own*—by the same wrongdoer on any item paid by the bank in good faith and without notice *after* the first item and statement are available to the customer for a reasonable period not exceeding fourteen days. 4–406(2)(b). Losses resulting from alterations and forgeries by the same wrongdoer on these later items usually can be traced directly to the negligent customer. If he had done his duty in the first place, the bank would have been alerted to stop payment on the later items and the wrongdoer would have been taken out of circulation or at least deprived of the opportunity to continue his misdeeds. Comment 3, 4–406.

Further justification for penalizing a customer who fails to discover and report his unauthorized signature or alteration with reasonable promptness is the fact that the bank's liability for paying forged or altered checks is, with few exceptions, absolute and without regard to fault on the bank's part. As long as the bank is free from fault, it is considered fair to protect it against the customer's carelessness.

However, if the customer can show that the bank, itself, failed to exercise ordinary care in paying an item, the customer is not precluded from asserting against his bank any unauthorized signature or alteration even though he has failed to exercise the required care. 4-406(3).

Even if the customer was not at fault and the bank was negligent in paying an item, the customer nevertheless is barred from asserting either *his own unauthorized signature* or an alteration unless he does so within one year after the statement charging it is made available to him; and he is barred from asserting that the signature of an *indorser* was unauthorized unless he does so within three years. 4-406(4). The time limits differ because of the assumption that there is far less excuse for a drawer not to detect an alteration or a forgery of his own signature than for failing to detect a forgery of an indorser's signature.

Example 83

Drawer drew a $5,000 check on Payor and issued it to Payee. Thief stole the check, forged Payee's signature and was paid by Payor upon presentment. Payor sent to Drawer a statement of accounts on May 1. Drawer failed to examine it until May 21 when Payee informed him that the check had been stolen. Drawer promptly notified Payor and demanded that the account be recredited. Payor claimed that Drawer had, by its delay and inattention, failed to exercise reasonable care and promptness under 4-406(1) and was precluded from asserting the forgery against Payor. Payor is wrong. In the absence of negligence by Drawer substantially contributing to the forgery, 3-406, Drawer has no duty under 4-406 with regard to

Ch. 13 CHECKS AND CHECK COLLECTION

a forged Payee indorsement. The duty runs to "his unauthorized signature." If, however, a second indorsement forgery by Thief occurs after the statement of accounts was available, Drawer would be precluded from asserting the forgery. 4–406(2)(b). In any case, Payor must have paid the item in "good faith" and the exercise of ordinary care. 4–406(3).

§ 8. Effect of Death or Incompetence

As a general rule, death or an adjudication of incompetency of his principal has the effect of terminating an agent's authority even before the agent learns of it.

As applied to the agency aspects of the bank-customer relationship, prior to the Code this rule was relaxed both by statute and by the cases. This relaxation of the general rule is reflected in the Code which provides that "neither death nor incompetence of a customer revokes * * * authority to accept, pay, collect or account until the bank *knows* of the fact of death or of an *adjudication* of incompetence *and has reasonable opportunity to act on it.*" 4–405(1). (Emphasis added.) The Code further liberalizes the rule by providing that "even *with knowledge* a bank may for ten days after the date of death pay or certify checks drawn on or prior to that date unless ordered to stop payment by a person claiming an interest in the account." (4–405(2). Emphasis added.)

One reason for not binding a bank until it actually knows of the death or adjudication of incompetency is that the tremendous volume of items handled by modern banks makes any rule requiring a bank to keep track of the life and competency of its customers very impractical. Comment 2, 4–405. As applied to paying checks, there is the further reason that a check is an order to pay which the bank must obey to avoid the risk of being held for a wrongful dishonor. Comment 2, 4–405.

The Code provision allowing a bank to pay or certify checks *even with knowledge* unless ordered not to is an innovation intended to permit holders of checks drawn shortly before death to cash them without the need for filing a claim in the estate proceedings. There is usually no reason why these checks should not be paid. Consequently requiring a holder to file a claim against the estate is a wasteful formality burdensome not only to the holder but also to the court, the bank, and the estate. Comment 3, 4–405. To provide for the unusual case where there actually is reason to question a holder's right to payment, the Code allows a relative, creditor, or any other person with an interest in the account to order the bank not to pay the check.

§ 9. Right to Stop Payment

It sometimes happens that a person who has issued a check discovers that he has been defrauded by the payee, that the check has been lost, that the check was issued by mistake, or that there is some other reason why he does not wish to have the check paid. He may try to protect himself by ordering his bank to stop payment. 4–403(1).

Although the stop payment device is often essential to the protection of the customer, it involves difficulty

Ch. 13 CHECKS AND CHECK COLLECTION

to the bank and it can easily be abused. Therefore, most banks charge for services connected with stop payment. Some banks charge only on special checking accounts and others charge only after a customer has repeatedly used the service.

To bind the bank, the stop payment order must be received by the bank in time to give it a reasonable opportunity to act before the bank has paid the check or has committed itself to recognize it by certifying or settling for it, or in any of several other ways. 4–403(1).

Because of the need for haste, most stop payment orders normally are given by telephone and later confirmed in writing. An oral stop payment order is binding upon the bank for only fourteen calendar days unless confirmed in writing within that period. A written order is good for six months unless renewed in writing. 4–403(2).

Only the bank's customer may stop payment; a payee or indorsee has no right to do so. An exception exists in case of the drawer's death when any person having an interest in the account may stop payment. Comment 3, 4–403, 4–405(2).

Since the drawer has no right to require a bank to impair its own credit, he has no right to stop payment of a check which has been certified. Even though his check has been certified, however, the drawer might accomplish the purpose of a stop payment order by promptly enjoining the payee's negotiation of the check and attaching the funds in the hands of the bank.

The effect of stopping payment is to hold the money in the drawer's account. It does not prevent the payee or other holder from suing the drawer on the check or on the obligation for which the check was given. Of course, if the drawer has a defense, he may assert it. Even though he has a defense, however, it will not protect him against a holder in due course who acquires the check unless it is one of the few defenses good against a holder in due course.

A payment in violation of an effective direction to stop payment is an improper payment and renders the bank liable for the loss resulting to the depositor even though it is made by mistake or inadvertence. Any agreement between the bank and customer relieving the bank of its duty to exercise good faith and ordinary care in such matters would be invalid. 4–103(1).

It is important to remember that a stop payment order creates a potential conflict with Payee and Collecting Bank on the one hand and creditors of Drawer, including Payor Bank, on the other. As we have seen, some of these priority conflicts are resolved under Section 4–303(1), which provides that a stop order, whether effective or not to terminate Payor's contractual duty to pay, "comes too late" if Payor has first done any of the acts specified. See Chapter 13, Section 2D.

§ 10. Payment of Stale Checks

In banking and commercial circles, a check outstanding more than six months usually is considered to be stale. Banks are not required to pay such checks, and many banks will not pay them without consulting its

Ch. 13 *CHECKS AND CHECK COLLECTION*

depositors. However, the bank is not required to consult the drawer, and if it pays an older check in good faith, it is entitled to charge the drawer's account. 4–404. This rule places the burden on the customer to give instructions to the bank regarding uncashed checks. Sometimes, as in the case of dividend checks, the drawer still wants payment made even after a considerable period of time.

§ 11. Bank's Right to Subrogation for Improper Payment

When a bank pays an item and charges the drawer's account despite a stop payment order or under other circumstances giving the drawer reason to object, justice between the bank and drawer may require: (1) recrediting the drawer's account with the full amount of the payment, (2) leaving the charge against the drawer, or (3) recrediting part of the amount paid. In some cases it also may be necessary to permit the bank to obtain some kind of relief against the party who obtained payment.

All of this is reflected in Section 4–407 of the Code which provides:

> If a payor bank has paid an item over the stop payment order of the drawer or maker or otherwise under circumstances giving a basis for objection by the drawer or maker, to prevent unjust enrichment and only to the extent necessary to prevent loss to the bank by reason of its payment of the item, the payor bank shall be subrogated to the rights
>
> > (a) of any holder in due course on the item against the drawer or maker; and

[*360*]

(b) of the payee or any other holder of the item against the drawer or maker either on the item or under the transaction out of which the item arose; and

(c) of the drawer or maker against the payee or any other holder of the item with respect to the transaction out of which the item arose.

In Universal C.I.T. v. Guarantee Bank & Trust Co., 161 F.Supp. 790 (D.C.Mass.1958), plaintiff, depositor, drew checks on defendant bank. The payee deposited these checks in Worcester Bank which allowed the payee to draw against them and so became a holder in due course. Learning that the payee was not entitled to payment, the depositor issued a stop payment order to defendant bank. By mistake, defendant bank paid the checks when they were presented by Worcester Bank and charged the depositor's account. When the depositor sued to recover the charges, the court held for the defendant bank on the theory that it was subrogated to the rights of the Worcester Bank which, as a holder in due course, would have been entitled to recover from the depositor even if the defendant bank had refused payment. This is consistent with Section 4–407(a).

Suppose that in the above case the payee had retained the check instead of negotiating it to a holder in due course and that the drawee bank had paid the payee by mistake despite the stop payment order. In this case, the drawee bank's right to charge the depositor would have depended upon what rights the payee had against the depositor. For example, if the payee already had performed all its obligations to the depositor and was entitled to full payment of the check, the bank, being subrogated to the payee's rights, would have been entitled to charge the drawer's account for the check's full amount. 4–407(b). If the payee had not carried out any part of the obligation for which the check was issued, the bank, as subrogee of the payee, would not have been entitled to charge the depositor's account at all. It would have been entitled, however, to recover from the payee on the theory that it was subrogated to the depositor's rights. 4–407(c). If the payee had performed only part of his obligation to the depositor, the bank would have been subrogated to the payee's claim and could have charged the depositor's account to that extent. It would have been entitled to recover the balance of the payment from the payee on the theory that it was subrogated to the depositor's claim against the payee. 4–407(c).

Of course, many cases are far more complicated. However, in all of these cases the right of subrogation is allowed only to the extent that is necessary to prevent loss to the bank and unjust enrichment to some other party.

Example 84

In payment for goods delivered, Drawer issued a check for $10,000 to Payee. Payee deposited the check in Depositary Bank which gave a provisional settlement for the full amount. Drawer then discovered that the goods were defective and issued a binding stop order under 4–403(1). In disregard of this order, Payor Bank paid the check upon presentment and debited Drawer's account. The payment was final and no presentment warranties were breached. Payor Bank refused to recredit the account and Drawer sued for damages under 4–403(3), alleging that but for the improper payment Drawer would have had a defense against and thus no obligation to Payee. Under the prevailing law, Payor Bank, whether it actually recredited the account or not, would have a valid defense under Section 4–407. Subrogation to Payee's rights against Drawer would be ineffectual because of the defense. See 4–407(b). But Collecting Bank is a holder in due course, 4–208, and Payor is subrogated to Collecting Bank's defense-free rights on the check against Drawer. 4–407(a). Drawer assumed the risk that the check would be negotiated to a HDC and subrogation puts the parties in the same position as if Payor had honored the stop payment order.

§ 12. Summary by Way of Example

A critical time in the check collection process is presentment. Should Payor Bank pay or dishonor the item? The answer turns on the contract of deposit with Drawer and whether the item is "otherwise properly payable" from the account. 4–401(1). If so, payment to the holder discharges liabilities on the instrument and the underlying obligation. If not, dishonor protects the interest of the Drawer and Payor Bank but does not otherwise affect the rights and duties of the parties. When "things go right" checks and the collection process facilitate the payment of obligations and "things go right" most of the time.

When "things go wrong" the matter becomes more complicated. Again, the trouble is manifested (but not necessarily caused) by Payor's decision upon presentment. The item may be improperly dishonored or improperly paid. Although both cause problems for

Ch. 13 CHECKS AND CHECK COLLECTION

Drawer and holders of the check, the latter is more frequent and the most difficult to unravel. Consequently, by way of summary we will examine again three common causes of trouble that produce improper payments by Payor: forged Drawer's signature; forged Payee's indorsement; and alteration. In each case there is a wrongdoer—call him Thief—who has received payment and disappeared, leaving others holding the bag.

A. FORGED DRAWER'S SIGNATURE

In a common situation, Thief steals blank checks, forges Drawer's signature and fills them in with Thief as payee. Drawer, of course, has no underlying obligation to Thief, the checks are never issued and there can *never* be a holder. Nevertheless, suppose Thief is able to persuade Grocer to "cash" the check and, upon presentment through the collection process, Payor Bank pays. A number of consequences flow from this improper payment.

First, Drawer has suffered a loss in the face amount of the check and Payor must recredit his account. Drawer did not "order" payment and the item, therefore, was not properly payable. 4–401(1).

Second, as against Drawer, subrogation under 4–407 does not help Payor here. There was neither a payee nor a holder that had rights against Drawer on the item.

Third, Payor Bank has made a final payment for purposes of 4–213(1). Further, let us assume that Collecting Bank has given Grocer a final settlement and

Grocer has changed his position in good faith in reliance upon the payment. See 3–418. Payor's only hope is to proceed against Collecting Bank and Grocer for breach of some warranty of presentment under 4–207(1). Thief, who might be liable on the instrument, 3–405, and, in any event, would be liable to Payor in restitution (read 3–418 again) is "long gone." But, alas, no warranties that all signatures were genuine were made or breached by either party. For historical and frequently debated reasons, see Comment 1, 3–418, the risk of payment over a forged drawer's signature is placed upon the bank, unless one or both presenters has knowledge of the forgery. 4–207(1)(b). What about the warranty of good title? The blank check was stolen. Yes, but it was never issued and was never an instrument. The "good title" warranty in 4–207(1)(a) refers to an "item" which is defined as an "instrument for the payment of money." 4–104(1)(g). So Payor Bank is stuck with the loss unless it has insurance or has made a prior agreement with Drawer.

Fourth, suppose Drawer elects not to have Payor recredit his account. Does Drawer have any claims against Grocer or Collecting Bank who dealt with the stolen check? Within Article 3, at least, the answer appears to be yes. Section 3–419(3) seems to permit Drawer to sue Collecting Bank in conversion to the extent of "any proceeds" of the check remaining in its hands. In this case there are none for a final settlement has been made with Grocer. But even so, some decisions have reasoned that 3–419(3) is inapplicable because a stolen blank check has no value and that in

Ch. 13 *CHECKS AND CHECK COLLECTION*

any event Drawer should be required to pursue the claim of improper payment against Payor. See, e. g., Stone & Webster Engineering Corp. v. First Nat'l Bank & Trust Co., 345 Mass. 1, 184 N.E.2d 358 (1962). Whatever Drawer's conversion claims are worth, Payor is not subrogated to them under Section 4–407(c) for there is no payee or holder in the transaction under discussion.

Fifth, Drawer's negligence, either in substantially contributing to the forgery, 3–406, or failing to examine any statement of accounts, 4–406(1), may preclude him from asserting the unauthorized signature against Payor. Drawer's negligence, which is a question of fact in each case, is a defense which must be raised and proved by Payor. If Payor was negligent in paying over the forged drawer signature, however, Drawer's negligence is neutralized and Payor must recredit the account. See 3–406 and 4–406(3).

B. FORGED INDORSEMENTS

In this case Drawer will owe Payee money and issue a check in payment. If the check is bearer paper, i. e., "pay to the order of cash," and it is stolen from Payee by Thief, the stage is set for some bad news for Payee. Although Thief is not a holder (the check was not delivered to him by Payee), the next person to whom Thief delivers the instrument and who takes in good faith, for value and without notice may be a holder in due course. Since Drawer ordered payment to bearer and a holder is a bearer, payment by Payor is not improper. Morever, payment to a holder discharges

Drawer from liability on the instrument and the underlying obligation. This leaves Payee holding the bag. If Thief is long gone, Payee has no claim against Drawer and neither Holder nor Payor have converted the instrument. Thus, the Code puts the loss on the holder from whom bearer paper was stolen whether he was negligent or not.

But suppose Drawer issued order paper, Thief stole it from payee and forged Payee's indorsement and delivered it to Grocer for value. Grocer, in turn, deposited it for collection and it was paid by Payor Bank. What are the legal consequences of this payment?

First, the payment is improper because Payor did not pay to Payee or order. Drawer's account has been debited in the face amount of the item and that would appear to be the loss caused by the improper payment. But remember that Drawer owes Payee the same amount which has not been paid. This complicates the analysis. Compare 4–403(3).

Second, as against Drawer, subrogation under 4–407(b) does help Payor. Even though Grocer cannot be a holder, see 3–202(1), there is a payee who has rights against Drawer. So, it would seem, Payor could defend an action by Drawer to recredit the account by claiming subrogation to Payee's rights. But Payor has paid Grocer not Payee. If the subrogation defense is successful, Drawer, who was not discharged, will still be liable to Payee. Perhaps we should take the analysis one step farther.

Third, in this case (unlike the forged drawer's signature), Payor does have a claim against Collecting Bank

Ch. 13 CHECKS AND CHECK COLLECTION

and Grocer for breach of a warranty that they have "good title to the item." 4–207(1)(a). The usual justification for protecting Payor here but not when the drawer's signature is forged is the difference in Payor's opportunity to verify the signature: Payor "is in a position to verify the drawer's signature by comparison with one in his hands, but has ordinarily no opportunity to verify an indorsement." Comment 3, 3–417. Grocer, on the other hand, may be in the best position to verify Thief's indorsement at the time the check is taken. In any event, warranty theory permits Payor to put the loss on Grocer, the first solvent person after the theft and forgery. Thus, Payor could recredit Drawer's account and be made whole against Grocer while Drawer would remain as before, liable to Payee on the underlying obligation.

Fourth, Payee, on the other hand, may sue Payor for conversion rather than Drawer on the underlying obligation. Section 3–419(1)(c) states that an instrument is "converted" when it is "paid on a forged indorsement." Section 3–419(2) provides that in an action against the drawee, the measure of liability is the "face amount of the instrument." Thus, if Payee is not negligent, see 3–406, and recovers the face amount of the instrument from Payor, that result in effect discharges Drawer. All of this brings us back to the 4–407 subrogation issue.

Fifth, in any event Payor is subrogated to Payee's rights to "prevent unjust enrichment and only to the extent necessary to prevent loss to the bank by reason of its payment of the item." If Payor has recovered in warranty from Grocer, then it has suffered no loss and, certainly, Drawer is not unjustly enriched. Drawer must still pay Payee for whatever was given in exchange. But if Payor must pay Payee the face amount in conversion, it has suffered a loss and Drawer has been enriched to the extent that he keeps goods or services delivered by Payee without payment. Subrogation under 4–407(b) seems proper there. The doubtful case for subrogation is where Payor refuses to recredit the account or pursue a warranty claim and makes no payment to Payee. If subrogation is available there, Drawer is liable twice in a case where Payor could have recovered in warranty and made no payment to the person to whose rights it is subrogated. Since Drawer played no part in the theft and forgery and is not unjustly enriched, it seems sound to deny subrogation and insist that Payor pursue its warranty claims.

C. ALTERATION

What is Payor's liability when an altered instrument is paid upon presentment? An important distinction must be made between a check issued with blanks and later filled up beyond authority and an instrument issued complete and later altered. In the former case, if Payor pays in good faith a holder it may charge Drawer's account "according to * * * the tenor of the completed item * * * unless the bank has notice that the completion was improper." 4–401(2)(b). In this case, the risk of a completion beyond authority is placed upon Drawer, who issued the check with blanks. In the latter case, Payor may charge Drawer's account "according to * * * the original tenor of his altered item."

Ch. 13 CHECKS AND CHECK COLLECTION

4–401(2)(a). In both cases, a Payor who pays a holder in good faith is given the same protection afforded a holder in due course of the altered item.

Suppose the check was issued for $500. Later it was stolen from Payee, skillfully raised to $600, cashed by Thief at Grocers and ultimately paid by Payor. Grocer, of course, is not a holder—the item was not negotiated to him. Thus, payment to Grocer would not discharge Drawer, 3–603, and apparently would not entitle Payor to charge Drawer's account in the amount of the original tenor of the altered item. Payment must be made in good faith "to a holder." 4–401(2). As discussed in the previous subsection, subrogation has its problems in cases of this sort. If Payor is subrogated to Payee's rights against Drawer without Payee having been paid, Drawer has a double liability without having been involved in the alteration or negligent. Again, the answer is found in warranty theory: Collecting Bank and Grocer warrant to Payor that the item has not been materially altered. 4–207(1)(c). Thus, Payor can recoup its loss from Grocer and Drawer, with a recredited account, can satisfy its obligation to Payee.

The case illustrates the importance of a definition. If holder is loosely defined, then Payor can charge Drawer's account in the amount of $500 and Drawer is discharged on the instrument. Payee, then, is made to bear the risk of the theft and alteration. One can question whether, without negligence, this result is sound. The risk, arguably, should be placed upon parties who took subsequent to the alteration because they were in the best position to verify indorsements and tenor. This result is achieved if a strict definition of holder is adhered to on these facts. But, as you can plainly see, there are questions of efficiency and fairness here presented that no definition can resolve. Until a more careful allocation of these risks among the parties is accomplished, we must work with what we have.

CHAPTER 14

BEYOND COMMERCIAL PAPER: ELECTRONIC FUND TRANSFER SYSTEMS

§ 1. Why a "Less-Check" Society?

Articles 3 and 4, developed in the early 1950's, assumed that paper—notes, drafts and checks—would be used in credit and most payment transactions. This assumption was firmly rooted in past and then current practice and was not shaken by the then imperfect view of the potential of electronic and computer technology. It is not likely that technology will ever fully replace paper in transactions where credit is extended. Some writing will be required to evidence the promise to pay with its surrounding terms and conditions. Where payment of the obligation is involved, however, the potential applications of technology are enormous. When currency is not used for payment, the source of "money" is the highly liquid demand deposits held by banks and other financial institutions. If the debtor's demand against these accounts and the bank's transfer to the creditor's account could be made electronically, the check as a means of payment could be eliminated. At the very least, electronic technology would produce a "less-check" if not a "check-less" society.

Why eliminate checks? There are two basic reasons, (1) excessive paper, and (2) the "float." Billions of checks are processed by banks and the Federal Reserve System each year. The system is virtually clogged with paper and any step which reduced the number of checks would be more efficient. The "float" describes the time between check issue and final payment when the drawer, in theory, still has use of the money. In some transactions, this could be 10 to 14 days. There may be as much as $6 billion in outstanding checks in the "float" on any given day. Thus, an electronic fund transfer system would give the payee-creditor immediate use of the money and eliminate an unintended credit feature from a payment transaction. It would insure prompt payment and make more efficient use of the demand deposits which economists call "M 1."

Despite its flexible structure and purposes, see 1–102(1), the Code is not well adapted to cover most electronic fund transfer systems. Thus, the American Law Institute established a subcommittee of the Permanent Editorial Board of the Code to consider whether Articles 3, 4 and 8 required amendment and, if so, in what manner. The so-called "3–4–8 Committee" has been at work for seven years and its report on Articles 3 and 4 is awaited with great interest. In the interim, Professors Penney and Baker have prepared an excellent study of the current status, problems and legal implications of the existing and proposed electronic technology. See N. Penney & D. Baker, The Law of Electronic Fund Transfer Systems (1980). At this time of transition, therefore, our job is to provide a few brief words of illustration and overview.

Ch. 14   *BEYOND COMMERCIAL PAPER*

§ 2. Check Handling Systems

A. MAGNETIC INK CHARACTER RECOGNITION (MICR)

The Magnetic Ink Character Recognition system (MICR) processes paper checks by electronic means. The checks are encoded with numbers which tell the electronic equipment (EDP) at the Depositary Bank how to sort the check for its proper destination and tell the Payor Bank whose account is to be charged. Also, the amount to be paid is encoded on the lower right corner, along with other symbols used for identification and control. Checks processed in this way are rarely examined manually. Fully in place since 1963, this mechanized system is a highly efficient method of processing but not necessarily reducing the paper flood of checks.

Some sections of Article 4 were drafted with electronic check-processing in mind. Section 4–204 permits an item to be sent directly to a Payor's automated processing facility rather than to Payor itself, Section 4–103 provides flexibility by agreement and the incorporation of Federal Reserve regulations and circulars as change occurs and Section 4–108 permits a collecting bank to give the Payor an extra day to handle an item in cases where, say, the electronic equipment has failed. And, as we have seen, Section 4–109 seems to anticipate electronic processing in the definition of "process of posting." See Chapter 12, Section 2.

A problem not clearly covered, however, involves an erroneous encoding on the check of the amount to be paid. Suppose the check is payable in the amount of $100 but is erroneously encoded in the amount of $1,000 by the depositary bank. This is not a material alteration because the contract of the drawer is not changed—the change is on the margins of the writing. See 3–407. But if the check is paid through the mechanized system in the amount of $1,000, the drawer will be out $900 and the payee will have an unexpected bonus. In cases of this sort, the correct outcome should be that the drawer's account is recredited and the payee must make restitution of $900 to the payor bank. This result is not clearly supported by Article 4, however, and so far there are no cases directly on point. See N. Penney & D. Baker at Par. 1.02. Compare 4–401, 4–406 and 4–207. This and other legal questions posed by breakdowns in the MICR system will, presumably, be covered in the final report of the "3–4–8" Committee.

B. CHECK TRUNCATION

Although the MICR system expedites the processing of checks, it does not eliminate them. Paid checks are duly returned with bank statements to customers, who then have a duty to examine them and report forgeries or other irregularities. See 4–406. A check truncation system, however, eliminates the return of the check from the payor to the customer. The payor retains and stores the paid check and sends, instead, a more detailed statement to the customer. In another planned but not operative system, the check, if paid, would be microfilmed and destroyed by the depositary bank. The banks would talk to each other through electronic information transfer and the payor would send a detailed statement to the customer.

If cancelled checks are not returned to the customer under a truncation system, what about the customer's duty to report forgeries under Section 4–406? If the check is not returned, this detection will be difficult if not impossible. Most commentators believe that the statement sent in lieu of a check will not satisfy any of the three conditions that trigger the customers duty to report under Section 4–406. A clear answer, therefore, will require an amendment to Section 4–406 or an agreement between Payor and its Customer that is responsive to the dilemma. Even there, an agreement which effectively disclaimed any Payor liability for forgeries in the guise of clarifying the Customer's duty to report them would probably go too far. See 4–103 and 1–102. See also, N. Penney and D. Baker, par. 2.02.

Example 85

Payor Bank adopted a check truncation system under which "paid" checks would be retained and an "Account Reconciliation" statement sent instead to Customer. The statement listed each payment by check number and amount, gave the date paid and sequenced the transactions by check number. If requested by Customer, Payor would immediately supply a photo copy of any check. Customer signed an amended deposit agreement providing that "failure of Customer to report any unauthorized signature within 30 days of receipt of the Account Reconciliation statement shall preclude Customer from asserting any unauthorized signature against Payor."

On June 1, a co-worker removed a check from Customer's checkbook, drew it in the amount of $100, entered his name as payee and forged Customer's signature as drawer. The thief also entered the number and amount of the check to "cash" in Customer's checkbook before returning it to her purse. Customer had no knowledge of these acts. Payor Bank cashed the check and stored it. On July 1, Payor sent Customer a statement for June which listed a sequence of 75 payments. Customer examined the statement, balanced her checkbook and found nothing amiss. On August 15, however, Customer did request a photocopy of the check from Payor Bank to resolve a recurring doubt about its origin. The forgery was discovered and Payor Bank refused to recredit the account on the ground that she was precluded from asserting the unauthorized signature.

Payor Bank is probably wrong. First, Customer arguably had no duty under Section 4–406(1) to exercise reasonable care and promptness to "examine the statement and items to discover his unauthorized signature." Since the item was not returned, the question is whether Payor made the item available to Customer in a "reasonable manner." If Customer could not reasonably detect the forgery from an examination of the statement, there could be no basis for requesting a photocopy, even though Payor Bank would supply it. The item, therefore, was not reasonably available. Similarly, if Customer cannot reasonably detect an unauthorized signature from the statement, imposing in the Deposit Agreement a 30 day limitation upon reporting forgeries is arguably unreasonable. Again, Customer should have from the statement a reasonable basis for determining the forgery and a reasonable time within which to report it before the agreed limitation is enforceable. See 4–403(1) and 1–102(3).

Ch. 14 *BEYOND COMMERCIAL PAPER*

§ 3. Electronic Fund Transfer Systems

Drawer, a California Corporation, owes Payee, a New York Corporation, $150,000 on a note. Payment is due. Drawer, by means of a computer located at corporate headquarters, orders Payor Bank (where Drawer maintains an account) to transfer $150,000 to Payee's account in a New York bank. Payor responds instantly to the order and transmits the credit, properly identified, to Payee's bank where it is properly entered. The entire transaction took 5 minutes. Drawer, an Ohio consumer, owes American Express, a New York Corporation, $123.46. Utilizing a home computer, Drawer orders Payor Bank to transfer the stated credit to American Express' account and the transfer is made electronically. Twenty five years ago this scenario might be rejected as ridiculous. Today, the electronic technology for such transfers is perfected. It is, according to some, just a matter of time before a pervasive system is in place which will eliminate both paper and the "float" in the payment process.

A. CURRENT SYSTEMS

Current systems in operation take steps toward but do not realize the electronic transfers noted above.

In larger transactions, it is possible for a Drawer-Customer to have the credit wired by Payor Bank to Payee's Bank through either Federal Reserve or alternate communication systems. These fund transfer messages, however, are not routine services provided in all transactions. In fact, the primary use of wire transfers is to pay and settle transactions between banks. Because the Customer must request and Payor Bank must agree to the wire in large transactions, its use in private payment is somewhat limited.

In high volume, low dollar transactions, banks use an Automated Clearing House to facilitate the clearing of checks presented for payment. The banks are members of the Automated Clearing House Association and the system is operated by the Federal Reserve. If the system is fully utilized, Depositary Bank, upon receipt of a check for collection, forwards the amount along with other items payable on a magnetic tape by courier to the ACH facility. The information is then processed by computer and a new tape is sent by courier to Payor Bank. In this process, the time in which debits and credits are entered and payments made is shortened. But checks must still be drawn and issued and returned. The basic system simply facilitates what might be called a paperless exchange of funds between banks.

The ACH system provides the structure within which certain pre-authorized transfers to and from individual accounts may occur. For example, if an employer and employee agree in advance, the employer can make a direct deposit to the employee's account as wages become due. The same service can be provided for other regular payments, such as dividends, pensions and social security. In a word, payment information is delivered by the employer to an originating bank, processed, sent on magnetic tapes or by wire to an ACH, processed again, and sent to the receiving bank, which credits the account. No checks are involved and, once the pre-authorization forms are signed, the transfers are automatic. The ACH can also be utilized for debit entries, with prior consent of the consumer. If a creditor, such as a mortgagee, initiates the procedure by sending billing information to the originating bank, the ACH system is then used to transfer a credit from the consumer's account to the creditor's. More recently, consumers have had some opportunity to initiate credit transfers through the ACH rather than having to depend upon pre-authorized, automatic transfers. This gives the consumer more control over the account, and of course, is a key step in any system where the customer can, without issuing a check, initiate an electronic transfer.

Two other automated systems are Automated Teller Machines (ATM) and Point of Sale Systems (POS). The former, now familiar in most urban areas, is an electronic funds transfer terminal that permits the customer to obtain information from and make transfers within his bank. Armed with a plastic card and a personal identification number, the customer can deposit and withdraw funds, transfer funds from savings to checking accounts, obtain short term loans, and the like. When operating, these machines save time and paper. When not operating * * * ah well. In a POS system, a purchaser, armed with proper plastic and identification, can effect an automatic transfer from his checking account to that of a seller to pay for goods purchased at the "point of sale." In short, the seller's computer talks to the computer in the buyer's bank and a credit is transferred. These systems are also used to capture and process data for the seller and to obtain check authorization or guarantees.

As demonstrated by this brief overview, the technology to accomplish the customer initiated automatic transfers posited at the beginning of this section is available but not in place. It is believed by most, that when the legal, political and economic impediments to a truly national system are overcome, the ACH will be the key facility in the electronic transfer.

B. LEGAL CONTROLS

The most comprehensive federal legislation is the Electronic Funds Transfer Act, enacted by Congress in 1978. 15 U.S.C.A. §§ 1693 et seq. (1979). The Act covers consumer fund transfers initiated through electronic equipment so as to "authorize a financial institution to debit or credit an account." Section 903(6). The Act excludes transactions initiated by check or draft, transfer by wire, certain internal transfers between Bank and consumer and the like. The Act deals with a wide range of problems, most of which are beyond the scope of this book. For a thorough analysis, see Penney and Baker, Par. 10.01 and Chapters 11–15. A particular problem, liability of the consumer for unauthorized transfers, deserves brief treatment.

Suppose Payor Bank debits consumer's account for a transfer which consumer claims was unauthorized. Consumer claims that her access card and PIN were stolen when a purse was taken on the subway and that $300 has been withdrawn through an automated teller from her account. She concedes that both the card

Ch. 14 *BEYOND COMMERCIAL PAPER*

and the PIN number, written on a 2 × 3 card, were in her billfold. In this case, Section 909(a) applies because the access card had been accepted by her *and* the issuer had also provided a PIN that must be used with the card to make a transfer. Otherwise, the full liability would be placed upon the issuer. On the facts as given, however, some liability for the unauthorized transfer is placed upon the holder of an accepted card that can't be used without a PIN if both access devices are stolen. The basis of liability is not stated as negligence, but the assumption is that the consumer is in the best position, once the card is accepted, to prevent theft of *both* card and PIN and to notify issuer of the loss.

What is the scope of liability? Section 909(a) states that it is the lesser of $50 or the amount of money obtained by the unauthorized use before the issuer is notified or becomes aware of the unauthorized use. Thus, if consumer had $1,000 in her account and the thief, in three separate transactions, withdrew $900 from the account before notice was given, issuer would bear $850 of the loss. There is one important exception, however. If issuer establishes that the loss was caused by consumer's failure to report the loss or theft of the card "within two business days" after learning of it, the consumer's liability is expanded to an amount not to exceed $500 or the amount of loss following two business days after the loss but before notice, whichever is less.

The complexity of this statute and the growing array of federal regulations around it is augmented by state statutes designed to accomplish roughly the same objectives. At this point in time, the line between federal and state regulation of electronic fund transfers and systems is not clear. Considerable care, therefore, must be taken in analyzing and assessing each transaction. See J. White and R. Summers, Uniform Commercial Code 16–9 (2d ed. 1980) for a more detailed discussion.

§ 4. A Final Word

The importance of the law of commercial paper in the business world should not blind us to the fact that these distinctive legal and financial devices are not ends in themselves but only means of achieving business objectives. To the businessman, the underlying business transaction is of primary importance; often much creative effort goes into it before any thought is given to its legal aspects. Even when attention is directed to the legal side of a transaction, the legal principles relating to commercial paper usually are no more important than those relating to contracts, agency, security, corporations, or other branches of the law that fix the guidelines within which business must be conducted. As desirable as it might be, however, it is not possible for any book to deal effectively with the law of commercial paper in terms of the multifaceted business transactions to which this law may relate. It follows that anyone aspiring to a sound understanding of the many legal and law-related aspects of business not only must grasp the principles of law of commercial paper but also must develop his capacity to apply these principles in a variety of transactions and to integrate them with the many other legal principles that may be applied. This is no easy task, but anyone genuinely interested in law and the business process will find the effort rewarding.

INDEX

References are to Pages

ACCELERATION CLAUSE
Effect on negotiability, 66
Notice instrument overdue, 256–257

ACCEPTANCE
See also, Certification
Banker's, 30
Effect of, 26–28
Extrinsic, 155-156
Presentment for, 188–189
Varying draft, 158
What constitutes, 26–27, 155–156

ACCEPTOR
Accrual of action, 170–171
Bank as, 30
Liability, 155–158, 183–184
Right to recoupment, 173–174

ACCOMMODATION PARTY
Accrual of action, 173
Liability, 164–167
Rights after payment, 164–167

ACCRUAL OF CAUSE OF ACTION
See Cause of Action

AGENCY
See also, Unauthorized Signatures
Effect unauthorized signatures, 145–146
Scope of authority to sign, 141–145

AGREEMENT
See also, Parol Evidence Rule
Defenses extrinsic to, 295–297

[*385*]

INDEX

References are to Pages

AGREEMENT—Continued
Immediate parties, 86–96
Interpretation, 87–88
Parol evidence rule, 88–96
Varying effect of Article 3, UCC, 11–13

ALLONGE, 121

ALTERATION
Customer's duty to report, 352–356
Effect of negligence, 149–151
Erroneous encoding not, 375
Liability on instrument, 146–148
Payor's liability, 369–371
Rights of holder in due course, 291-292
Warranties against, 214–215, 229–232

ALTERNATIVE DRAWEES
Effect on negotiability, 55

ALTERNATIVE PAYEES
Effect on negotiability, 71

AMERICAN LAW INSTITUTE, 5

ANTEDATING
Effect of negotiability, 80–82

APPROPRIATE PARTY TO INDORSE, 102–104
See also, Indorsement

ARTICLE 3, UCC
Agreement varying effect, 11–13
Drafting history, 7–8
Exclusions from scope, 10–12
Relationship to Article 9, UCC, 12
Relationship to Code, 34–35
Relationship to Uniform Consumer Credit Code, 280–281
Scope, 8–13

ARTICLE 4, UCC
Adaptibility to change, 374–375
Relationship to Articles 3 and 9, UCC, 10
Scope, 328–331

INDEX

References are to Pages

ARTICLE 9, UCC
Effect on assignment, 41–43
Relationship to Articles 3 and 4, UCC, 10–12
Relationship to Code, 34–35

ASSIGNEE
See Assignment

ASSIGNMENT
Assignee's risk, 38–40
Common law development, 37–40
Effect in consumer transaction, 270–282
Effect of Article 9, UCC, 41–43
Effect of negotiability contrasted, 43–47
Effect of "shelter" principle, 268–269
Issue of check as, 154–155, 343
Negotiation contrasted, 99–100
Partial negotiation as, 120
Warranties of assignor, 39–40

BANKS
See also Collecting Bank; Payor
Payor bank and customer, 341–371
Role in check collection, 328–341

BEARER PAPER
How negotiated, 104–106
Order paper distinguished, 101
Risks of, 106–108

BILL OF EXCHANGE
See also, Draft
Early law, 3–4

BLANK INDORSEMENT
See Indorsements

BURDEN OF PROOF
Establishing,
    Claims and defenses, 302–303
    Rights against third parties, 303–304
    Rights on instrument, 300–302
    Status as holder in due course, 303

INDEX

References are to Pages

CANCELLATION
As discharge, 325

CAUSE OF ACTION
Accrual of, 169–174

CERTIFICATE OF DEPOSIT
Form and parties, 21–22
Form and uses, 32–33

CERTIFICATION
    See also, Acceptance
Acceptance by bank, 30–31
Effect, 343–345
What constitutes, 155–158

CHECK COLLECTION
    See also, Checks
Check handling systems, 373–375
Collecting bank as holder in due course, 269–270
Customer's rights, 332–341
Liability on improper payment, 363–371
Outside banking system, 126–129
Process described, 125–126, 328–331
Restrictive indorsements, 125–135
"3–4–8" Committee, 373
Warranties in, 235–236

CHECKS
    See also, Check Collection
Cashier's check, 20–21
Certification by drawee, 343–345
Collection process, 328–331
Form and parties, 18–21
Form and uses, 31–32
"Less check" society, 372–373
Traveler's check, 21

CLAIMS
Against holder in due course, 298–299
Against person not a holder in due course, 299
Agreement not to assert, 268–269
Effect of notice, 251–264

INDEX

References are to Pages

CODE
See Uniform Commercial Code; Uniform Consumer Credit Code

COLLECTING BANK
　See also, Banks
As agent for collection, 332
As holder in due course, 269–270
Defined, 328
Duties in check collection, 332–333
Presentment by, 193

COMMERCIAL PAPER
Doctrinal lynchpins, 36–37
Early law, 3–4
Forms, 14–22
　Certificate of deposit, 32–33
　Check, 31–32
　Draft, 26–30
　Promissory note, 23–26
Importance, 1–3
Negotiation overview, 43–47
Uses, 23–33

CONDITIONS
Conditional delivery, 93–95
Determined by instrument, 55–57
Effect of another writing, 57–58
Effect on negotiability, 56–57
Implied not recognized, 58–59
Reference to fund or account, 60–61
Words negating negotiability, 59–60

CONFESSION OF JUDGMENT
Effect on negotiability, 78

CONSIDERATION
　See also, Value
Lack of as defense, 293–295
Value compared, 239–248

CONSUMER PROTECTION
Justification for, 10–11, 270–272, 281–282

INDEX

References are to Pages

CONSUMER PROTECTION—Continued
Prohibiting "cut off" devices,
    Courts, 272–276
    Federal Trade Commission, 279–282
    Uniform Consumer Credit Code, 276–279

CONTRACT
Between customer and payor bank, 341–345
Between immediate parties to instrument, 86–96

CONTRACT RIGHTS
Assignment at Common Law, 37–40
Effect of Article 9, UCC, 41–43

CONTRIBUTION
Accommodation party, 164–167

CONVERSION
Drawee's liability for, 155
Liability for, 163

CUSTOMER
    See also, Drawer
Account charged, 347–351
Contract with payor, 341–345
Duty to examine statement, 352–356, 375–376
Effect of death or incompetence, 356–357
Rights,
    Electronic funds transfer, 381–383
    In check collection process, 332–341
    To stop payment, 357–359
    Under Article 4, UCC, 328–331
    Upon improper payment, 363–371

COUNTERCLAIMS
    See also, Claims
Effect of notice, 251–264

DEATH
Effect in check collection, 356–357

DEFENSES
Discharge as, 312

INDEX

References are to Pages

DEFENSES—Continued
Effect,
    Against holder in due course, 283–292
    Against non-holder in due course, 293–297
    Agreement not to assert, 268–269
    Of notice, 251–264
    Of warranty, 215–216
Holder vs. assignee, 99–100
Jus tertii, 297–298

DEFINITE TIME
Requisite of negotiability, 66–69

DELIVERY
Conditional delivery, 93–95
Effect of, 86
First delivery as issue, 96–98
Meaning of, 85
Non-delivery as defense, 295
Subsequent delivery, 96–97

DEMAND INSTRUMENT
When payable, 65–66

DEPOSITARY BANK
Defined, 328
Restrictive indorsements outside banking process, 126–129
Role in check collection, 329–331
Supplying missing indorsements, 118–119

DISCHARGE
By,
    Acceptance varying draft, 158
    Alteration of instrument, 146–148
    Cancellation, 325
    Impairment of recourse or collateral, 321–325
    Payment or satisfaction, 313–318
    Reacquisition, 319–320
    Renunciation, 326–327
    Tender of payment, 318–319
Defense against holder in due course, 290–291
Drawer's secondary liability, 179–182
Effect of restrictive indorsement, 135–136

INDEX

References are to Pages

DISCHARGE—Continued
Indorser's secondary liability, 176–179
Liability on instrument, 311–327
Party with no recourse, 320–321
Underlying obligation, 306–311

DISCLAIMER
Qualified indorsement, 122–123

DISHONOR
Condition to secondary liability, 176–184
Giving notice of, 197–200
Liability for wrongful dishonor, 345–347
What constitutes, 195–197

DOMICILED INSTRUMENT, 193–195

DRAFT
See also, Checks
Acceptance varying, 158
Banker's acceptance, 30
Form and parties, 17–18
Forms and uses,
    Collecting accounts, 26–28
    Documentary transactions, 28–30
Trade acceptance, 29–30

DRAWEE
See also, Payor Bank
Defined, 17–19
Liability, 153–155, 183
Multiple, 55

DRAWER
See also, Customer
Accrual of cause of action, 171–172
Liability,
    Conditions, 179–182
    Excuse of conditions, 203–207
    On instrument, 152–153
Maker contrasted, 17
Named as drawee, 54

INDEX

References are to Pages

DURESS
Defense against holder in due course, 287

ELECTRONIC FUND TRANSFERS
Current systems, 378–381
Legal controls, 381–383
"3–4–8" Committee study, 8

EVIDENCE
Notice of dishonor, 197–198

EQUITY
Supplemental principles, 11

EXECUTORY PROMISE
Notice of defense or claim, 242–243
Value, 241–242, 243–245

EXONERATION
Accommodation party, 166

EXTENSION CLAUSE
Effect on negotiability, 67–68

FEDERAL RESERVE SYSTEM
Role in check collection, 330
Role in electronic fund transfers, 378–381

FEDERAL TRADE COMMISSION
Role in consumer transactions, 279–282

FICTITIOUS PAYEES
Problem, 108–113

FIDUCIARY
Restrictive indorsement to, 133–135

FINALITY
Payment,
    Under Article 3, UCC, 220–222
    Under Article 4, UCC, 335–341

FOREIGN BILL
When protest required, 201

INDEX

References are to Pages

FORGERY
See Unauthorized Signature

FORMS
Forms of commercial paper, 14–22

FRAUD
Defense against holder in due course, 288–290
Fictitious payee, 108–113
Imposters, 113–116

GOOD FAITH
Consumer transactions, 273–274
Requisite for holder in due course, 249–251
What constitutes, 249–251

GUARANTEE
What constitutes, 167–169

GUARANTOR
Accrual of cause of action, 172–173
Liability, 167–169

HISTORY
Assignments at common law, 37–40
Code predecessors, 5
Commercial paper, 3–4
Drafting Uniform Commercial Code, 5–8

HOLDER
Appropriate party to indorse, 103–104
Defined, 103
Owner contrasted, 106
Payment to as discharge, 313–314
Rights on instrument, 139–140, 300–301
Warranty claim, 209–210

HOLDER IN DUE COURSE
Assignee with advantages of, 268–269
Claims against, 298–299
Collecting bank as, 269–270
Consumer transactions, 270–282
Defenses against,
    Arising at or after transfer, 284–292

INDEX

References are to Pages

HOLDER IN DUE COURSE—Continued
Arising before negotiation, 284–292
Effect of,
    Alteration, 147–148
    First delivery, 96–98
    Warranty, 217–218, 233
Excluded transactions, 267–268
Importance, 237
Overview, 43–47
Payee as, 264–266
Requisites, 36–37, 238–264
"Shelter" principle, 266–267

ILLEGALITY
Defense against holder in due course, 287–288
Effect on negotiability, 78–79

IMPOSTERS
Problem, 113–116

INCAPACITY
Defense against holder in due course, 286

INDORSEMENTS
Appropriate party, 102–104
Blank and special compared, 101–102
Effect of limiting words, 136
Effect of restriction on discharge, 135–136
Importance in negotiation, 101–102
Missing,
    Depositary bank's power to supply, 118–119
Place on instrument, 120–121
Qualified and unqualified compared, 122–123, 159–162
Restrictive,
    Conditional, 124–125
    For deposit or collection, 125–135
    Prohibit further transfer, 124
Trust indorsements, 133–135

INDORSER
Accural of cause of action, 171–172

INDEX
References are to Pages

INDORSER—Continued
Liability,
    Among indorsers, 159–161
    Conditions to, 176–179
    Effect qualified indorsement, 161–162
    Excuse of conditions, 203–207
    Unqualified indorsement, 159–161

INFANCY
Defense against holder in due course, 285–286

INSOLVENCY
Defense against holder in due course, 290
Effect of warranty, 216–217

INSTRUMENT
    See also, Liability on Instrument
Contract between immediate parties, 86–96
Declaration of negotiability, 74–75
Domiciled, 193–195
Effect incompleteness, 80
Holder in due course must take, 238
Liability on, 137–174
Lost or stolen, 300–301
Place of indorsement, 120–121
Presumed negotiable, 82
Requisites for negotiability, 49–74
Transactional overview, 43–47
Within scope Article 3, UCC, 8–10

INTEREST
Rate to be charged, 173–174

INTERMEDIARY BANK
Effect restrictive indorsement on, 129–132

ISSUE
Discharge of underlying obligation, 306–311
Meaning, 83–85
Transactional overview, 43–47

JOINT PAYEES
See Multiple Payees

INDEX

References are to Pages

JUS TERTII
As defense, 297–298

LIABILITY ON INSTRUMENT
See also, Instrument
Conditions to secondary liability, 175–207
Effect of qualified indorsement, 122–123
Effect of restrictive indorsement, 135–136
Excuse of conditions, 203–207
In general,
    Accrual of cause of action, 169–174
    General principles, 138–151
    Particular parties, 151–163
    Special kinds, 163–169
    Various bases, 137–138
When presentment necessary, 185–195

MAKER
Accrual of cause of action, 170–171
Liability on instrument, 152
Nature of primary liability, 183–184

MONEY
Instrument payable in, 64–65
Sum certain, 62–64

MULTIPLE PAYEES
Effect on negotiability, 71
Requisites for negotiation, 119–120

NEGLIGENCE
Check truncation systems, 375–377
Customer's duty to payor, 352–356
Forgery or alteration, 149–151

NEGOTIABILITY
Assignment contrasted, 43–47
Concept illustrated, 9–10
Effect of,
    Another writing, 57–58
    Date, 80–82
    Declaration of, 74–75
    Implied conditions, 58–59

INDEX

References are to Pages

NEGOTIABILITY—Continued
    Missing words of, 75–76
    Risk allocation, 36–37, 47–48
Requisites of, 49–82
Sum certain, 62–64
Policy justification, 47–48
Terms and omissions not affecting, 77–79
Transactional overview, 43–47
When promise or order conditional, 55–57

NEGOTIABLE INSTRUMENT
See Instrument

NEGOTIATION
Assignment contrasted, 99–100
By multiple payees, 119–120
Effect if incomplete, 116–117
Effect if rescindable, 121–122
Fictitious payees, 108–113
Imposters, 113–116
Less than balance due, 120
Misspelled or mistated name, 121
Requisites, 100–106
Transactional overview, 43–47

NOTICE
    See also, Holder in Due Course; Notice of Dishonor
Defense or claim, 258–264
Dishonor, 197–200
Effect of omission or delay, 203–207
Instrument overdue or dishonored, 254–258
Multiple claims and defenses, 264
What constitutes, 251–254
When holder has, 254–264

NOTICE OF DISHONOR
    See also, Notice
Condition to secondary liability, 176–184

ORDER
Requisite of negotiability, 53–54

INDEX

References are to Pages

ORDER PAPER
Bearer paper distinguished, 101
Risks of, 106–108

OWNER OF INSTRUMENT
Holder contrasted, 106

PAROL EVIDENCE RULE
Scope and effect, 88–96

PAYABLE ON DEMAND
Requisite for negotiability, 65–66

PAYABLE TO BEARER
Requisite of negotiability, 72–73

PAYABLE TO ORDER
Requisite of negotiability, 69–71

PAYEE
As holder in due course, 264–266
Defined, 15
Fictitious payees, 108–113
Negotiation by multiple payees, 119–120

PAYMENT
As discharge, 313–318
Effect stop order, 357–359
Presentment for, 189–191
Stale checks, 359–360
Subrogation after improper, 360–363
Tender of, 318–319
When final, 220–222, 335–341

PAYOR BANK
Charging customer's account,
    Amount to be charged, 350–351
    Overdraft, 347–348
    Payment to non-holder, 348–350
    Unauthorized signature, 348
Contract with customer, 341–345

INDEX

References are to Pages

PAYOR BANK—Continued
Customer's,
    Death or incompetence, 356–357
    Negligence as defense, 352–356
    Right to stop payment, 357–359
Defined, 328
Effect restrictive indorsement, 129–132
Liability,
    Improper payment, 363–371
    Payment stale check, 359–360
    Unauthorized fund transfer, 381–383
    Wrongful dishonor, 345–347
Subrogation rights, 360–363

POSTDATING
Effect on negotiability, 80–82

PRESENTMENT
Effect of omission or delay, 203–207
Time of, 188–195
What constitutes, 185–188
When required, 176–184

PROCEDURE
Establishing claims and defenses, 299–305

PROMISE
Express v. implied, 52–53
Requisite of negotiability, 52–53

PROMISSORY NOTE
Early law, 4
Evidence of previous debt, 26
Form and parties, 14–17
Uses,
    To borrow money, 23–25
    To buy on credit, 25–26

PROTEST
Effect of omission or delay, 203–207
How made, 201–203
When required, 200–201

INDEX

References are to Pages

RATIFICATION
Effect on unauthorized signature, 146

RECOUPMENT
By indorser against maker or drawer, 177–179

REIMBURSEMENT
Accommodation party, 164–167

REMEDIES
Breach of warranty, 233–235
Subrogation, 360–363
Wrongful dishonor, 345–347

REMITTER
Defined, 103
Not a holder, 103

RESTITUTION
For payment by mistake, 220–221

RESTRICTIVE INDORSEMENT
See Indorsements

RENUNCIATION
As discharge, 326–327

RESCISSION
Effect on negotiation, 121–122

SATISFACTION
As discharge, 313–318

SETTLEMENT
  See also, Payment
Provisional settlement under Article 4, UCC, 333–335

SECONDARY LIABILITY
See Liability on Instrument

SET–OFF
Assignee subject to, 38–39
Notice of, 252–253

INDEX

References are to Pages

SIGNATURE
See also, Unauthorized Signature
Authorized, 141–145
Burden of establishing, 301–302
Capacity in which signed, 140–141
Essential for negotiability, 52
Holder need not sign, 139–140
Must appear on instrument, 138–139
Negligence contributing to forgery, 149–151
Requisites of, 140
Unauthorized, 145–146
Warranty,
    Authorized, 229–232
    Genuine, 212–214

SPECIAL INDORSEMENT
See Indorsements

STOP PAYMENT ORDER
See Customer; Payor Bank

SUBROGATION
Remedy of payor bank, 360–363
Right of accommodation party, 166

SUM CERTAIN
See Negotiability

SURETY
See Accommodation Party

TIME
Notice of dishonor, 199–200
Presentment, 188–195
Protest, 201

TITLE
Warranty of, 210–212, 222–225

TRADE ACCEPTANCE
Time draft, 29–30

INDEX

References are to Pages

TRANSFER
See also, Assignment; Negotiation
Discharge of underlying obligation, 306–311
Indorsement prohibiting, 124

TRANSFEREE OF INSTRUMENT
Right to indorsement, 116–117
"Shelter" principle, 266–267

UNAUTHORIZED SIGNATURE
See also, Signature
Customer's duty to report, 352–356
Drawer's signature, 364–366
Indorsements, 366–369
When defense against holder in due course, 292

UNDERLYING OBLIGATION
Discharged by issue or transfer, 306–311
Suspended upon issue, 306–311

UNIFORM COMMERCIAL CODE
Coverage, 6–7
Drafting history, 5–8
Overlaps among Articles, 34–35

UNIFORM CONSUMER CREDIT CODE
Eliminates "cut-off" devices, 276–279
Overlap with Article 3, UCC, 280–281

UNIFORM NEGOTIABLE INSTRUMENTS LAW
Purpose and effect, 5

UNQUALIFIED INDORSEMENT
See Indorsements

VALUE
Accommodation party, 164–165
Bank credit as, 245–247
Executory promise as, 241–245
Holder in due course must give, 239–248
Security for antecedent debt, 247–248
Lack of consideration, 293–295

WARRANTY
Assignment, 39–40

INDEX

References are to Pages

WARRANTY—Continued
Check collections, 235–236
Defenses, 215–216
Effect of final payment, 220–222
Effect on holder in due course, 217–218, 233
Insolvency proceedings, 216–217
Liability,
    Presenters, 218–233
    Transferors, 208–218
Material alteration, 214–215, 229–232
Remedies for breach, 233–235
Signatures, 212–214, 225–229
Title, 210–212, 222–225

WITHOUT RECOURSE
See Indorsements

WRITING
Effect on negotiability, 57–58
Requisite of negotiability, 51

WRONGFUL DISHONOR
See Customer; Dishonor; Payor Bank

†